LIFE IN A SHELL

LIFE IN A SHELL

A PHYSIOLOGIST'S VIEW
OF A TURTLE

Donald C. Jackson

HARVARD UNIVERSITY PRESS

Cambridge, Massachusetts

London, England

First Harvard University Press paperback edition, 2013

Library of Congress Cataloging-in-Publication Data

Jackson, Donald C., 1937–

   Life in a shell : a physiologist's view of a turtle / Donald C. Jackson.

     p. cm.

   Includes bibliographical references and index.

   ISBN 978-0-674-05034-1 (cloth : alk. paper)

   ISBN 978-0-674-07230-5 (pbk.)

   1. Turtles—Physiology.   2. Animals—Physiology.   3. Physiology,
Comparative.   I. Title.

   QL666.C5J28 2010

   571.1792—dc22

                                      2010018567

To Diana, Tobey, and Thomas

CONTENTS

1 The Turtle's Shell   1

2 Buoyancy   19

3 The Breathing Turtle   33

4 Tortuguero   53

5 Overwintering without Breathing   70

6 Living without Oxygen   89

7 The Heart of a Turtle   113

8 Life in the Slow Lane   135

Epilogue   150

References   159
Acknowledgments   171
Index   173

LIFE IN A SHELL

1

THE TURTLE'S SHELL

A turtle lives inside a shell. This is a familiar fact that we all take for granted, but what does it mean to live this way? A shell is heavy and cumbersome. How does it affect a turtle's ability to move about, on land or in water? Because the shell is so heavy, what keeps an aquatic turtle from sinking to the bottom? How does a turtle breathe without an expandable rib cage like the one we have? How do turtles manage to have sex with intervening armor? These and many other questions arise when you begin to think seriously about life in a shell. These are the kind of questions that biologists ask themselves and perhaps even try to find answers to. Some of these questions, and others as well, have occupied me during my career as an animal physiologist. My purpose in writing this book is to tell how I and others have studied these questions and what we have discovered about life inside a shell.

I did not grow up dreaming of becoming a turtle biologist. In fact, it was not until I finished graduate school and was doing postdoctoral research that a chance circumstance led me to study this animal. Even then, I thought of it as temporary, in order to complete a particular project. Yet the results of that study, as results of studies tend to do, raised more questions than they answered, so when I set up my own laboratory I decided to start out by studying turtles. Like a snowball rolling down a hill, my fascination with turtles grew, and soon I was hooked. Looking back, I realize I could easily have chosen a different animal— but I chose a turtle.

I now know that this was a wise choice. The turtle's various special features have provided a lifetime of opportunity for discovery, for me and for many other biologists, and I know that many more lifetimes could still be profitably spent probing ever more deeply into the mysteries of this animal's biology. Yet I also realize that however special the turtle is, each organism has novel features that enable it to adapt to its particular environmental niche. If you chose an organism at random and devoted yourself to its study, then you would probably not be disappointed. So much can be learned about an individual organism: its population biology, the nature of the environment in which it lives and how it has adapted to live and thrive in that environment, its behavior, its anatomy, its organismal and cellular physiology and biochemistry, and its molecular biology—all of these contribute to defining an organism and understanding its place in the natural world. Yet for almost all living organisms, with the exception of ourselves, common laboratory animals, and a few model species, our understanding is comparatively meager.

Some practical considerations prevent the scientific community from achieving a broad understanding of all living things, the first being the sheer number of individual species that exist out there. If each would require many lifetimes of study, then too few scientific lifetimes are available. In addition, if a researcher wishes to obtain financial support to study an organism, then an important scientific objective is required other than to simply learn what you can about a random creature. My studies of turtles almost always were directed at basic principles that could be studied easily in turtles or perhaps were expressed in dramatic fashion in these animals. Understanding how something works in a model animal can often help us understand that same function in other animals, including humans. A major focus of my research has been devoted to how certain freshwater turtles can survive for long periods in the absence of oxygen. One particular turtle, the painted turtle, is probably the most highly adapted vertebrate recognized in this regard. As I discuss later in this book, a painted turtle can survive several months without oxygen during winter hibernation. For this reason, if you are interested in knowing how a vertebrate can live without oxygen, then the painted turtle is an ideal animal to study.

This remarkable trait of the painted turtle is an illustration of what animal physiologists call the "Krogh Principle." August Krogh, a Danish physiologist of the early twentieth century, won the Nobel Prize in 1923 for his work on capillary exchange, but he was a biologist with broad interests in many aspects of animal physiology and medicine. In a paper based on his talk at the International Physiological Congress, Krogh (1929) wrote: "For such a large number of problems there will be some animal of choice or a few such animals on which it can be most conveniently studied" (247). These lines define what is now called the Krogh Principle. For example, if you wish to study the ionic basis of action potentials in nerve, then an ideal animal is the squid, with its giant nerve axon; if you want to study the impact of hydrostatic pressure on the cardiovascular system, then an ideal animal is the giraffe, with its long neck; and if you wish to study survival without oxygen, then the painted turtle is a logical choice.

Turtle Myths

Turtles have long held an important place in the popular imagination. The turtle's strength, persistence, patience, and slow, deliberate movements have made it an iconic figure and given it a prominent role in the myths and legends of many cultures. One variation of a familiar Native American creation story tells of an island in the sky where people lived and a world below covered in water where only aquatic animals lived. A young woman from the sky island falls through a hole and plummets toward the water but is caught and saved by two birds that place her, for safety, on the back of a giant turtle that has swum to the surface. After some deliberation, a frog dives far down to the bottom, brings back mud to the surface, and places the mud on the turtle's back. By some mysterious process, the mud grows and grows until it becomes a continent. The young woman brings forth children, and these children bring forth still more children, and so habitable land is formed and populated with the ancestors of the Native Americans, all of these events supported on the sturdy shell of the giant and faithful turtle. In a remarkably similar worldview from the Hindu tradition, the earth rests on the back of a giant turtle, often with one or more elephants in between the earth and the turtle. What supports the giant turtle is not clear, although in some accounts the turtle is swimming through space patiently bearing its load. Many variations of these myths exist, and it has been suggested that some among us still believe these versions of "reality." I read the following, unattributed, possibly apocryphal story some years ago. Other versions have also been published.

> After delivering a lecture on the solar system, philosopher-psychologist William James was approached by an elderly lady who claimed she had a theory superior to the one described by him.
>
> "We don't live on a ball rotating around the sun," she said. "We live on a crust of earth on the back of a giant turtle."
>
> Not wishing to demolish this absurd argument with the massive scientific evidence at his command, James decided to dissuade his opponent gently.

3

"If your theory is correct, Madam, what does this turtle stand on?"

"You're a very clever man, Mr. James, and that's a good question, but I can answer that. The first turtle stands on the back of second, far larger, turtle."

"But what does this second turtle stand on?" James asked patiently.

The old lady crowed triumphantly, "It's no use, Mr. James—it's turtles all the way down."

The most familiar story in our culture involving a turtle is Aesop's fable about the tortoise and the hare (Figure 1.1). The slow but steady tortoise wins the race, an apt metaphor for the image of a turtle slowly paddling through the water, plodding across a meadow, or patiently waiting for spring at the bottom of a frozen pond, never in a great hurry but eventually reaching its goal. Although Aesop himself could not have anticipated a broader interpretation, the fable can also celebrate the evolutionary longevity of the turtle, for this reptile began its journey a long time ago, well before we (or the hare) appeared on earth, and it has watched many life-forms drop out of the race as it has journeyed on.

Evolution and Development

The turtle's evolutionary history stretches back to the late Triassic period, over 200 million years ago, when the world was a very different place than the one we know today. Global temperatures were much warmer, and the North and South poles were ice free and mild. All landmasses were combined in a single supercontinent called Pangaea, where much of the vegetation and animal life would appear quite strange to a time traveler today. Yet if that visitor happened upon a creature lumbering across that long-ago landscape the animal would look reassuringly familiar, for extending from within its oval, rigid shell were a head and neck at the front, a tail at the back, and sprawling limbs at the sides. This time traveler may have encountered one of the earliest-known turtles, now called *Proganochelys*. Some of its features differed from modern turtles—its tail contained spikes, with a club at the end, its neck had protective spines, and you probably would not want to try to pick up this turtle, which was approximately 1 m in length (Gaffney, 1990). Yet this ancient animal was still unmistakably a turtle, its iconic shell being the giveaway (Figure 1.2).

Figure 1.1 The race between the tortoise and the hare (Milo Winter from Aesop, 1984).

Like shells of contemporary turtles, the shell of *Proganochelys* consisted of plates of bone enveloping the internal organs and incorporating the ribs and the central portion of the spinal column. Also as in modern turtles, the rib cage was fused to the inside of the upper shell, and the shell itself surrounded the pectoral (or shoulder) girdle, a structural trait of turtles that was and continues to be unique among the vertebrates (Figure 1.3). This body plan has served the turtle well and has remained

Figure 1.2 Model of the ancient turtle *Proganochelys* (from Gaffney, 1990). Courtesy The American Museum of Natural History.

essentially unchanged through the thousands of millennia since the Triassic period. These inscrutable survivors have watched the Age of Dinosaurs come and go, they have watched the rise of mammals, and quite recently they have seen the emergence and explosive increase of humankind. Since they evolved, turtles have survived major extinction events, including the Triassic-Jurassic event at the end of the Triassic period 200 million years ago and the Cretaceous-Tertiary (K-T) event at the end of the Cretaceous period 65 million years ago, when 50% of all species disappeared. Now the biosphere is in the midst of a major new extinction event, this one largely of humankind's own making. All species are at risk, including our own. Many species of turtles are threatened because of loss of habitat and human exploitation. Can the turtle once again survive? I sincerely hope so. Perhaps even when we are gone, the turtle will still be here, wrapped in its protective shell, continuing its slow, persistent journey.

Fossil records will not likely provide a complete record of an animal's evolutionary history, and for many years *Proganochelys,* a fully formed turtle, was the oldest known turtle. Fossils of this turtle appeared abruptly in the late Triassic period, with no obvious intermediate forms preceding it. However, this has now changed with the discovery of a new fossil in China, dated to 220 million years ago, making it some 10 million years older than *Proganochelys* (Li et al., 2008). This new species has been given

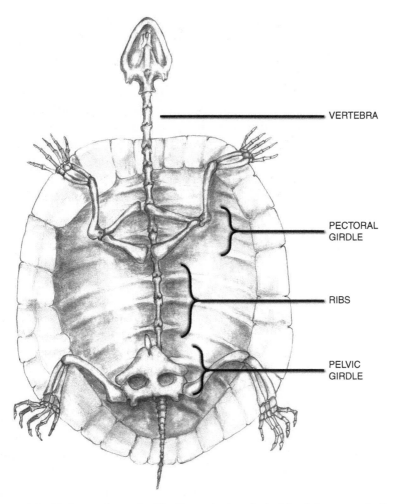

VERTEBRA

PECTORAL
GIRDLE

RIBS

PELVIC
GIRDLE

Figure 1.3   Skeleton of a turtle. The figure depicts the inside of the upper shell, the carapace, with the embedded ribs and the overlying pectoral and pelvic girdles. The position of the ribs outside the pectoral girdle is unique to turtles.

the Latin name *Odontochelys semitestacea* (try saying *that* fast three times in a row). This discovery is exciting and significant because it appears to be an earlier stage in the evolution of turtles, a kind of missing link. *Odontochelys* has a fully formed plastron (bottom shell), but its carapace (upper shell) is poorly developed and consists only of expanded ribs and some bony plates along the spinal cord. Major changes in our

7

understanding of the turtle's evolution have been prompted by this discovery. For one, the plastron must have evolved prior to the carapace, a sequence that agrees with the embryological development in modern turtles. Also, based on where this fossil was found, Li et al. (2008) believe that *Odontochelys* was primarily aquatic. This challenges the prior theory, based on the primacy of *Proganochelys,* that the original turtles were terrestrial. This discovery, therefore, challenges many prevailing ideas about the evolution of turtles, including the identity of the stem group from which they arose.

Of course, it is possible to interpret a fossil discovery in different ways, and this is no exception. An alternative interpretation is that *Odontochelys* was itself derived from an earlier terrestrial turtle that had a full shell (Reisz and Head, 2008). The loss of the bone on its carapace, according to this view, was due to its adoption of an aquatic habitat. Some aquatic turtles that live today, such as the softshell, are secondarily derived in this fashion. Of course, this game of second-guessing can be played forever, since we can never know for certain when, finally, the earliest turtle has been found. Occam's razor, the rule that favors the simplest explanation for something, in this case should award most primitive status to the oldest known turtle, *Odontochelys.* Without stronger counterarguments, the safest conclusion is that at this point in the evolution of turtles a full plastron was present but a bony, mineralized carapace had not yet evolved. This earliest turtle also had a full complement of teeth, in contrast to *Proganochelys*, which had teeth only on the roof of the mouth, and to modern turtles, which lack teeth entirely and instead have a horny beak. These distinctive features are the basis for its Latin name, *Odontochelys semitestacea,* translated as "toothed turtle with a half shell."

Our latest oldest-known turtle, *Odontochelys,* did have a carapace, but it was not hard and bony and was therefore not preserved in the fossil record. What survived were the bony ribs that provide the general shape of the carapace. In modern turtles, and in *Proganochelys,* these ribs are fused into the bone of the carapace, a particular type of bone called dermal bone because it is derived from the skin. *Odontochelys* would have looked like a turtle, but if you pressed on its top shell it would probably have felt more leathery, like a softshell turtle. As in other turtles, its ribs extended like long slender arms arching outward toward the lateral margins of the shell, unlike the ribs of other vertebrates that curve downward and then inward toward the midline to form a cage around the upper-body cavity. Based on studies of snapping turtles, this critical difference in the trajectory of the growth of the ribs occurs early in embryological development and is the point during development when a turtle stops looking like any other vertebrate and starts looking like a turtle. The ribs of the turtle embryo sweep upward and to the sides and are incorporated into the overlying dermis. The ribs eventually terminate at the left and right sides at a site unique to turtles called the carapacial ridge (Burke, 1989). The carapacial ridge is thought to release chemical factors that attract the growing ribs into the overlying dermis and so form the framework of the overarching carapace (Cebra-Thomas et al., 2005), although recent work suggests instead that the carapacial ridge is responsible for the fanlike distribution of the ribs (Nagashima et al., 2009). Eventually, bone, dermal bone, forms at sites along the ribs, and these sites expand and coalesce to complete the formation of the rigid carapace. In the process of this novel growth trajectory, the carapace has enclosed the shoulder girdle within its structure. Earlier, dermal bone had also formed on the ventral (front or belly side) of the turtle, producing the bottom part of the shell, called the plastron. A bridge of bone on each side connects the plastron to the carapace, thus forming the complete shell. We can now suppose that the development of *Odontochelys* terminated prior to the ossification of the carapace, and that this further ossification was a later step in the evolution of turtles as we now know them.

The evolutionary origin of turtles has undergone a recent reexamination. It has long been thought, based on skull anatomy, that turtles are the only surviving members of an ancient line of reptiles called anapsids. The term *anapsid* refers to the absence of openings, or fenestrae, in the temporal region of the skull. It was thought, therefore, that turtles derived from a separate evolutionary path than other reptiles called diapsids that have two openings on each side in the temporal region. Recent molecular evidence, however, has challenged this traditional idea and indicates instead that turtles share a common ancestor with a cluster of organisms called archosaurs, which includes birds and crocodilians, and that this ancestor in turn shared a still-earlier ancestor with snakes and lizards (Zardoya and Meyer, 1998; Iwabi et al., 2005). According to this view, turtles descended from diapsid forebears but secondarily lost the openings in their skulls and became anapsid. It is worth noting that dinosaurs were on the archosaur line leading to birds, so turtles share a common ancestor with dinosaurs as well.

The Turtle's Shell

Whether you are a scientist interested in a turtle's evolution or development or physiology, a folklorist interested in the turtle's place in literature or legend, or just someone who likes to observe turtles in nature or keep one as a pet, the central feature that captures your attention is the shell. This is what sets the turtle apart from its fellow vertebrates, influences so much of its physiology and behavior, and makes it instantly identifiable to everyone. A turtle on land must carry this heavy load when it travels, certainly a major handicap in its race with the hare. Yet because of its strength, the shell can support a great weight (the whole world?). The practical importance of the shell for its owner is self-evident and has played a critical role in the turtle's evolutionary success story. As surrounding armor, the shell has afforded a formidable defense against external dangers of both the animate and inanimate varieties. Even outrageous contemporary images of turtles, such as the Teenage Mutant Ninja Turtles, the sewer-dwelling superheroes that wield martial arts weapons with humanlike hands while gritting their full sets of teeth, retain some authenticity by having shells that make them recognizable as turtles and of course employing them as protective armor against blows from evildoers. Most turtles do not possess a ninja turtle's weaponry, so they rely more on the retraction of head and limbs within their shell if escape to a safe haven is impossible. A turtle's habit of retreating into its shell suggests that it is timid and withdrawn, not willing to face the world. We apply this to people we know who are shy, and we encourage them "to come out of their shell." But a turtle's retreat into its shell is not a behavioral flaw potentially correctable by appropriate counseling; rather, it is a strikingly successful defensive strategy that has served it well through the ages.

Shell Structure

What is the shell, how does it compare to the bones of other animals, and what other functions does it serve for the turtle inside? A turtle's shell is a combination of two types of bone: dermal, which is formed by ossification within the dermis of the skin, and endochondrial, or endoskeletal, which is bone that begins its life as cartilage and then is ossified. In human skeletons, the vertebrae, ribs, limb bones, and jaw are endochondrial bone, whereas the flat portions of the skull and clavicle (collarbone) are dermal bone. Extensive dermal bone occurs in many other vertebrates, both living and extinct. Familiar dermal bone in living animals includes the bony bumps called osteoderms distributed on the backs of crocodiles and alligators, the scales of fish, and the armor plates of armadillos. Dramatic examples of dermal bone also are apparent in fossils of extinct animals. The plates and spikes on dinosaurs such as the genus *Stegosaurus* were dermal bone, as was the armor of ancient fish and amphibians. Thus the presence of generous amounts of dermal bone is not unique to turtles; indeed, it has been suggested (at least it was until our new friend, *Odontochelys,* was discovered) that turtles evolved from armored creatures called pareiasaurs. What *is* special is the incorporation of the vertebrae and ribs into that bone and the enclosure of the limb girdles within the ribs.

Most turtles, however, do have an exceptionally large amount of bone compared to other vertebrates, and they are particularly well endowed with dermal bone. In the familiar freshwater species, the painted turtle, the shell accounts for about 32% of the total body weight of the animal. In addition, the skeleton outside of the shell, which includes the shoulder and hip girdles, the limb bones, and the skull, adds another 5.5%, so the total bone mass approaches 40% of the animal's weight. In comparison, the South American crocodilian, the caiman, has an estimated bone mass, including its skeleton and osteoderms, that is only about 14% of its body mass (Jackson et al., 2003).

The shell may seem like an inert suit of armor, but in reality it is a living part of the turtle. Like vertebrate bone generally, the turtle's shell has living cells, nerves and blood vessels, and grows and remodels. The turtle's shell, like its skeletal bone, also closely resembles all vertebrate bone in its composition. Much of the shell consists of crystals of calcium phosphate held within a matrix made up of a protein called collagen. Like other bone, the shell also contains additional elements and molecules, some in rather large amounts. For example, in the painted turtle, more than 99.9% of the total body calcium and more than 98% of the phosphorus are located in its shell and skeleton. These structures also contain over 90% of the magnesium, over 60% of the sodium, and over 95% of the carbon dioxide (mostly in the form of carbonate) contained in the turtle's body.

Shell Functions

The most obvious function of the turtle's shell is protection. The heart, the lungs, and other viscera reside safely within the protective armor and are less subject to injury than in soft-bodied animals, even those with rib cages. Although the head and appendages can project from the shell so that the turtle can engage in activities such as swimming, walking, breathing, mating, and feeding, most turtles other than marine turtles can retract these structures into the shell, and in extreme examples, such as the familiar box turtle, can tightly seal everything inside using hinged portions of the plastron. *Proganochelys*, incidentally, was apparently unable to retract its head and limbs, but its spiky, clublike tail may have provided an offensive weapon to deter potential predators. Most modern turtles emphasize a defensive approach to external threats, with a notable exception being the snapping turtle, with its powerful jaws and reduced relative shell mass. The two major groups of living turtles, Pleurodires, found only in the Southern Hemisphere, and Cryptodires, residents of both hemispheres and the majority of living turtles, including marine turtles, differ in how they retract their head and neck into the shell. A Pleurodire, or side-necked turtle, rotates its neck laterally so that the head and neck lie snugly against the upper part of the body. A Cryptodire, in contrast, pulls its neck straight back, and as the neck folds, the head is brought safely within the shell. The muscle that retracts the neck in a Cryptodire, the *retrahens capitis*, is a large and powerful muscle, as anyone who has tried to forcibly pull a turtle's head out of its shell realizes. In both groups of turtles, this strong, defensive maneuver of retraction of the head and neck into the enclosing rigid shell has provided an effective refuge from external danger.

The shape and composition of shells also differ among turtles because of competing considerations in shell design. Strength and resistance to crushing are probably paramount in terrestrial turtles such as land tortoises and box turtles (Figure 1.4). These species have rather heavy, dome-shaped shells. Besides making it more difficult for a predator to enclose the shell in its jaws, the domed shell, with its greater curvature, is a stronger design mechanically and is hard to crush. Think of the greater difficulty in crushing a chicken egg by squeezing on its domed ends versus squeezing it in the less curved middle portion. Aquatic turtles such as painted turtles (Figure 1.4) have flatter shells and thereby sacrifice some of the shell's strength in the interests of streamlining for easier swimming. Softshell turtles take this to an extreme by flattening the shell even further and greatly reducing the amount of dermal bone, and this is thought by some to enable them to wiggle their way into the mud at the bottom of a stream (Figure 1.4). Their shells provide only modest protection from predators, but they have sacrificed strength for flexibility in

BOX TURTLE

PAINTED TURTLE

SOFTSHELL TURTLE

Figure 1.4   Three types of North American turtles with contrasting shells.

order to exploit their effective sanctuary in the mud, where they are difficult to detect.

Protection is not the only function of the turtle's shell. The shell's fixed surface also serves as a site for the attachment of muscles. Various muscles, such as the pectoralis major and deltoid, both of which help control movement of the forelimbs, extend from the plastron to the limb bones. Contraction of these muscles contributes to the turtle's limb movements when swimming and walking. The openings of the shell at the front and back are covered with typical reptilian nonmineralized skin that is securely attached to the inner surfaces of the margins of the carapace and plastron and is continuous with the skin covering the head, the limbs, and the tail. The turtle is thus fully enclosed by typical reptilian skin or by mineralized skin, the dermal bone of the shell. The shell in turn is covered by keratinized scutes that impart the characteristic pattern and color to the turtle's shell.

The shell also serves as the major mineral reserve of the turtle's body. As noted earlier, the shell and skeleton contain the bulk of the body's calcium, phosphorus, magnesium, and sodium. Most importantly, these elements can move out of the shell into the blood and back into the shell from the blood and can thereby contribute to the regulation of their concentrations in the blood and other body fluids. In addition, the shell and skeleton can release certain of these elements to meet special needs of the turtle. Egg-producing females release calcium and phosphate in response to the hormone estrogen, and these minerals attach to a blood protein, vitelline, which then transports them to the oviduct, where they are used to make the eggshell (Dessauer, 1970). Mineral release also occurs when an aquatic turtle remains submerged for a long period of time with little or no available oxygen and produces large amounts of lactic acid, the chemical by-product of anaerobic (no oxygen) metabolism. Calcium and magnesium emerge from the shell into the blood in combination with carbonate to help buffer the lactic acid, similar in principle to an antacid tablet we swallow to neutralize excess stomach acid. A detailed discussion of this important function, central to the extraordinary ability of freshwater turtles to endure a lack of oxygen, occurs later in this book.

Although all turtles share basic features that identify them as turtles, considerable diversity in form and function enables them to occupy a multitude of ecological niches. By one count (Ernst and Barbour, 1989), worldwide there are some 257 living species of turtles. Some are principally aquatic, including both freshwater and saltwater species, some live in deserts, and some are tropical, while the range of other species extends into high latitudes with long, cold winters. As adults, turtles range in size from species such as mud turtles, with a shell length of 8-9 cm which weigh only about 100 g, to the giant leatherback sea turtle, which can reach a length of 2 m and a weight of 900 kg. Remarkably, however, a newborn leatherback hatchling, at 40–50 g, is smaller than the diminutive mud turtle.

The term *turtle* is often used generally to refer to all members of the subgroup (order) of reptiles called *Testudines,* although it is often used to refer specifically to the aquatic species. Land turtles are called tortoises, although in the United States this term is restricted to members of the family *Testudinidae,* which includes the desert tortoise and gopher tortoise. The term *terrapin* is sometimes used to describe freshwater turtles, although again in the United States this name is reserved for the diamond-back terrapin, *Malaclemys terrapin,* a brackish water species found along the East Coast. A recent phylogeny of turtles is shown in Figure 1.5, which depicts the ancient split between Pleurodires and Crytodires and the subsequent division into the thirteen families of living turtles. The common names of the various species referred to in this book are shown to the right side of the figure next to their family affiliations.

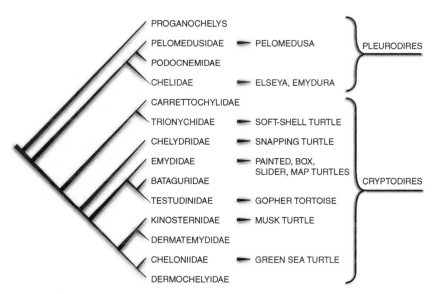

Figure 1.5 Turtle phylogeny (adapted from Shaffer et al., 1997) depicts the course of the evolution of turtles.

Goals and Issues

In the following chapters, I discuss the physiology of turtles. This book is not, however, a comprehensive discussion of all turtle physiology but principally concerns those topics that have occupied my attention over the years. As will be evident in these discussions, the shell significantly influences many aspects of a turtle's physiology and behavior. Both freshwater turtles and marine turtles are able to compensate for their heavy shells when in water and can move with little obvious effort through the water at close to neutral buoyancy (Chapter 2). On land, however, the shell makes travel more difficult, as is particularly evident in a nesting sea turtle hauling itself laboriously across the sand (Chapter 4). Because the act of breathing is constrained by the shell, turtles employ a different mechanism than occurs in other vertebrates, yet freshwater species can accomplish this important function with modest effort, even when they greatly increase their breathing (Chapter 3). One of the shell's positive contributions is as a barrier between the turtle and the outside world, but for a turtle spending the winter in a frozen stream, this barrier can restrict its uptake of oxygen from the water that can otherwise help it survive in this state (Chapter 5). For turtles spending the winter in ponds with little dissolved oxygen, however, the shell serves a crucial function by releasing buffers to neutralize metabolic acids (Chapter 6). Much of a turtle's physiology is not directly influenced by the shell. The responses of a turtle's regulatory system to a change in temperature (Chapter 3), the metabolic changes that occur during submergence (Chapter 6), the specialized features of the cardiovascular system (Chapter 7), and many other internal functions of a turtle do not directly relate to the shell. Nevertheless, even these functions have special properties that make turtles worthy topics for study.

Some of the material presented in this book may be challenging for some readers because of scientific complexity, although I have made every attempt to make the ideas as accessible as possible. For the more advanced reader who may be disappointed by oversimplifications, I have included references to primary sources where the details can be found.

For many of the experiments described in this book, it was necessary to humanely kill animals. Physiological research that attempts to understand the inner workings of organisms frequently requires the collection and analysis of tissues and cells. The number of animals used was always the least amount required to achieve the goals of the study, and in fact the minimization of animal usage is an important part of the governmental regulations that concern animal research. The National Institutes of Health and the U.S. Department of Agriculture have established guidelines to ensure that scientists who utilize animals in their research programs not only restrict the number of animals used but also minimize animals' pain and distress and also to ensure that the animals are housed in appropriate facilities. Research institutions and universities are required to have Animal Care and Use Committees that oversee animal research to make certain that these regulations are followed. My research was always conducted in accordance with these guidelines.

The use of animals in research is a serious matter, thus the governmental principles also require the research to contribute to the advancement of knowledge, the improvement of human or animal health, or the good of society. As stated earlier, my research always concerned basic physiological questions that could best be studied in turtles or that revealed some significant new information about this animal. A number of the topics that I dealt with have clear relevance to human physiology and medicine. For example, the ability of the painted turtle to live for months at a low temperature without oxygen identifies this animal as an ideal organism for understanding vertebrate adaptations to lack of oxygen, known as hypoxia. Hypoxia is a critical factor in many common medical conditions, such as heart attack and stroke. In addition, studies of turtles, including work in my laboratory, revealed the important way in which cold-blooded animals regulate their blood pH when body temperature changes, a finding that has direct relevance to the management of patients undergoing hypothermic surgery. The crucial role that the turtle's shell and skeleton play in neutralizing acid produced during winter hibernation is a dramatic example of how bone can function as a source of buffering, again an important finding relevant to human medicine.

To ensure that animal research satisfies the criterion of fundamental significance, peer review of research projects occurs both within an investigator's institution and at the agencies or foundations that fund the research, such as the National Science Foundation, the major funding source for my research. Trivial or wasteful projects can be prevented from being carried out by these processes. The species of freshwater turtles that were the subjects of most of my studies, the western painted turtle *(Chrysemys picta)* and red-eared slider *(Trachemys scripta elegans),* are probably the most readily available and most studied species of turtles. These are not threatened species, although I acknowledge that this consideration does not make the life of an individual turtle any less precious.

I consider it a great privilege to have had the career I did in physiological research. I owe much to the excellent mentors and role models that I had throughout my career, many of whom are mentioned in this book. I am also indebted to the many collaborators and students who have worked with me and contributed greatly to the work discussed here. And, above all, I pay tribute to the turtles themselves, and I sincerely hope that what I say in this book will enhance even further the reader's admiration and understanding of these extraordinary animals and contribute in some way to the turtles' protection.

2

I used to keep turtles in my lab in a large glass aquarium. They were red-eared sliders (scientific name: *Trachemys scripta elegans*), a species commonly sold in pet stores and also a common species in turtle research. Often I watched the turtles swim lazily about or remain motionless in mid-water and then periodically paddle up to the surface to breathe. Usually my mind would remain blank, hypnotized by the effortless and graceful movements of these creatures that were clearly at home in their natural element. As I watched, though, I gradually grew curious about how they were able to hang suspended in water, neither sinking nor floating, matched perfectly to their watery environment. Turtles after all have a shell that is much heavier than water and the shell must make the turtle tend to sink to the bottom. How, I wondered, does the turtle overcome that tendency to sink? How, in other words, does the turtle maintain neutral buoyancy despite its shell and bone and other tissues that have specific gravity greater than water?

Buoyancy Control in Freshwater Turtles

I decided to perform a simple experiment. I taped a small piece of lead to the underside (the plastron) of one of the turtles. Lead is more than eleven times as heavy as water, and if placed in water it will sink rapidly to the bottom. I selected a turtle that was approximately 500 g in weight and taped onto it a piece of lead weighing about 25–30 g. I returned the turtle to the tank and watched it sink steadily to the bottom. It sat there looking, to my eye, somewhat puzzled. It was late in the afternoon and time to go, so I took one last look at the turtle sitting on the bottom of the tank, with its neck starting to extend upward, before I turned out the light and left for the day. Now this seemingly coldhearted act of mine was tempered by the fact, well known to me at the time because of on-going experiments in my lab, that a turtle can easily survive a night sitting on the bottom of a tank without breathing. The turtle would not necessarily be happy about it but could do so. Plus I was confident that the turtle's swimming muscles would enable the animal to overcome this rather modest load on its shell and rise to the surface to breathe.

The next morning I arrived at my lab with considerable anticipation to learn how the turtle had managed overnight. To my astonishment, the turtle was swimming about the tank as effortlessly as ever, with the lead weight still attached to its shell. I removed the lead weight, returned the turtle to its tank, and was only mildly surprised when it bobbed to the surface. It was evident that sometime after I had departed the evening before the turtle had swum to the surface, probably with some effort, and eventually had filled its lungs with enough air to counterbalance the lead weight. This restored its neutral buoyancy and permitted normal activity in the tank. When the weight was removed, the overinflation of its lungs caused the turtle to float at the top.

Inspired by this observation, I set about designing an experiment to describe the mechanism the turtle employed (Jackson, 1969). My single experiment was already, in a sense, conclusive, but proper scientific practice requires multiple repetitions of the experiment on a sufficient number of animals to provide convincing, statistically validated evidence for your conclusions. In addition, the overall mechanism may not be as simple as it first appeared. Thus I had to test a number of turtles with lead weights taped to their shells, and for comparison I also tested the same turtles with Styrofoam floats attached to their backs. Styrofoam has quite low specific gravity (<<1) and tends to make the turtle rise to the surface.

My experimental design was as follows: I determined the turtle's initial specific gravity by first weighing it in air and then weighing it while it was completely underwater. To weigh the turtle underwater I placed it in a preweighed wire cage suspended from the below-table hook of a top-loading electronic balance. Specific gravity was calculated by dividing the animal's weight in air by the difference between its weight in air and its weight in water. The initial specific gravities of the turtles averaged about 1.05, slightly heavier than water (specific gravity = 1.0). Next I taped either a lead weight or a Styrofoam float to the turtle's shell, producing, respectively, an increase in specific gravity to more than 1.1 or a decrease in specific gravity to less than 1.0. I then returned the turtle to the tank, waited about a day, and then repeated the measurement of specific gravity. Consistent with the original observation, the turtles' (a total of eight animals, which were all tested with both attachments) buoyancy returned to normal, due to approximately equal changes in their weights in both air and water. Both weights increased in response to floats and both decreased in response to weights.

To confirm that the change in weight in water was due to a change in lung volume, I measured the lung volume of thirteen turtles that were exposed to a range of weights and floats. I used a method called plethysmography to do so. This involved submerging a turtle in a rigid, sealed chamber with no air and measuring the pressure change that resulted from injecting a small volume of water into the chamber. Assuming that lung gas is the only compressible substance in the chamber, the volume of that gas could be calculated using Boyle's law. The results revealed a linear and negative relationship between lung volume and specific gravity; that is, as lung volume decreased, specific gravity (not including the weight and floats) increased, and, conversely, as lung volume increased, specific gravity decreased. Remarkably, the turtles were able to reset their lung volume, and by that I mean the stable volume of their lungs between breathing episodes, over a very large range, from less than 3% of their body volume to almost 17% of their body volume. I was also able to establish that the neutral buoyancy of a turtle with no attachments is achieved with a lung volume of about 14% of body volume. By comparison, our resting lung volume is only about 5%–6% of our body volume, but then we do not have a shell.

The unanticipated observation of the experimental turtles was the weight change in air that was in the same direction and of similar magnitude to the weight change in water. Although lung gas has a large effect on weight in water (1 mL of gas provides a buoyant force of close to 1 g), gas has no effect on the animal's weight in air. The weight change in air must have been due to a movement of water in or out of the animal. The transfer of water could have occurred either through the turtle's mouth or through the opening in its tail. As a further experiment, I chose to alter the exchange through the tail by securing a soft rubber plug into the cloaca, the chamber inside of the tail. Cloaca is the Latin word for sewer, and in reptiles the cloaca is the common final pathway out of the body for products of the digestive, urinary, and reproductive systems.

I repeated the experiment with weights and floats on an additional eight animals, each with its cloaca blocked to prevent water movement. I observed that the turtles' normal buoyancy was restored in response to the floats, but that they could not compensate for the lead weights. These results provided insight into the mechanism the turtle employs. In response to a Styrofoam float, the turtle decreased its lung volume and took in water to keep its total body volume about the same. These changes produced increases in the turtle's weight both in water and air. With the cloaca occluded, the turtle could still add water by drinking and was not handicapped. On the other hand, to regulate its buoyancy in response to an added lead weight, the turtle had to increase its lung volume and to expel an equal volume of water. The pathway for water expulsion must be via the cloaca, because when the cloaca was occluded, buoyancy correction did not occur.

Additional experiments led me to conclude that a turtle can move water in and out of its cloaca using the same pumping mechanism that it uses to breathe air into and out of its lungs. The contraction of the respiratory muscles acts by expanding and contracting the soft limb pockets at the front and back of the animal and produces pressure changes that are uniform throughout the body cavity. Whether the turtle breathes air or water depends on which avenue is open. For lung ventilation, the turtle, with its head out of water, opens its glottis and air passes in and out of its lungs; for cloaca irrigation while submerged, the turtle opens its anal sphincter and water passes in and out of its cloaca.

Role of the Cloacal Bursae

Cloacal "breathing" in many species of turtles involves not just the cloaca but also specialized structures that open off each side of the anterior wall of the cloaca, called cloacal bursae, also known as cloacal or accessory bladders (Figure 2.1). Cloacal bursae are thin-walled sacs found in some, but not all, freshwater turtles and are absent in marine turtles. Their function has been debated for many years. Suggested functions include water storage, salt uptake, and oxygen uptake from the water. Water storage may be important for female turtles during nesting. Females of all species of turtles, even the most aquatic ones, return to land to lay their eggs. Archie Carr, in his classic book on turtles (1952), cited earlier observations by Cagle on red-eared sliders nesting in Tennessee: "He [Cagle] found that the dry soil in which the turtles were nesting was moistened prior to excavation by large amounts of water from the cloacal bladder, resulting often in the wetting of an area two feet or so in diameter" (256).

Moistening the nest cavity, however, is not a function for the cloacal bursae that would be useful in male turtles, although the structures are

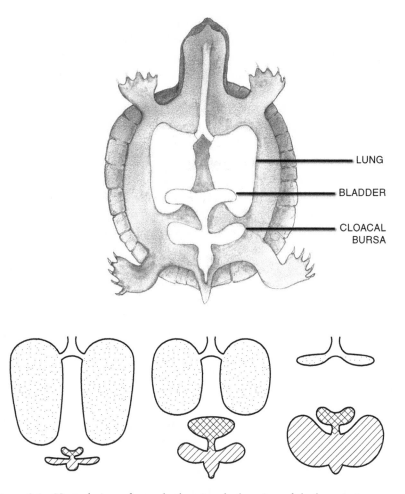

LUNG

BLADDER

CLOACAL
BURSA

Figure 2.1   Ventral view of a turtle showing the location of the lung, urinary bladder, and cloacal bursae (above) and how the volumes of these structures change during buoyancy adjustments (below).

present in both males and females. Water storage could also benefit turtles as a source of water during times when they are on land losing water through evaporation. The urinary bladders of turtles, as well as amphibians such as frogs, are known to serve this role. However, observations by Peterson and Greenshields (2001) indicate that turtles do not "drink" with their cloacal bursae, even when significantly dehydrated. Because freshwater turtles spend most of their time in ponds or streams where the water is much more dilute than their body fluids, dehydration is usually less of a problem than water uptake and salt loss.

Another possible function of the cloacal bursae is salt uptake from the surrounding water. Experiments by William Dunson (1967) revealed that some freshwater turtles, including *Trachemys scripta,* can transfer sodium ions from the water into their body across several surfaces, including the cloacal bursae, a process that can help maintain salt balance in a dilute environment. Still another suggested function of the cloacal bursae is respiratory gas exchange, in particular oxygen uptake from the water. This function is decidedly true for some Australian turtle species such as the Fitzroy River turtle *(Rheodytes leukops).* This unusual turtle has highly vascularized cloacal bursae that present a very large surface for exchange, analogous to the gills of fish. As described by Gordos et al. (2003), *Rheodytes* spends much of its time resting on the bottom of a stream pumping water in and out of its cloaca and rarely comes up to breathe. It can derive most of the oxygen it requires by diffusion from the water across the vascularized wall of the cloacal bursae.

My observations on buoyancy control indicated that the cloacal bursae can also serve as ballast tanks, such as are employed by submarines. Turtles, after all, resemble little submarines because of their rigid shell and relatively fixed volume. As in a submarine, a change in the volume of one of a turtle's bodily components, such as gas volume, must be balanced by an opposite change in another component, such as water, in order to avoid a large change in internal pressure. Unlike submarines, though, turtles can expand and contract their total volume because their rigid armor is interrupted at sites where their head and limbs emerge. Volume change is essential, of course, in order for the turtle to breathe. However, a persistent distortion of volume associated with a buoyancy adjustment will produce unacceptable changes in internal pressure, just as in the submarine. However, by reciprocal changes in water volume in or out of the cloacal bursae, the turtle can regulate its total body volume and maintain its internal pressure in response to steady-state displacements of lung volume (Jackson, 1971b).

I tested the volume-compensating role of the cloacal bursae in a more direct way by conducting the following experiment. I placed a turtle in a pure oxygen environment for one hour until its breathing had washed out all of the nitrogen from its lungs. I then submerged the turtle in a sealed, rigid chamber, a plethysmograph, with no access to air for about two hours. During this time, I measured the turtle's lung volume, as described earlier, by periodically injecting a small volume of water into the chamber and recording the resulting pressure change. I found that lung volume fell steadily as the lung oxygen was consumed by the turtle's metabolism; after two hours, the lungs were nearly collapsed. Collapse occurred this rapidly because nitrogen, an inert gas physiologically, was not present to keep the lungs inflated. After opening the chamber, I removed the turtle and without delay held it in a head-up position with an empty glass beaker positioned beneath its tail. As the turtle breathed and refilled its lungs, water squirted out of its cloaca into the beaker, and when no additional water was forthcoming, the final volume of water in the beaker was nearly identical to the decrease in lung volume that I had measured during submergence. Clearly, as the lungs gradually collapsed during submergence, water entered the cloaca to replace the lost volume. To confirm that the water actually entered the cloacal bursae, another turtle was submerged for two hours after breathing oxygen but was then euthanized without being permitted to breathe. Upon dissection of this animal, I found that the cloacal bursae were greatly distended with water.

Another Advantage of Large Lungs

In addition to counterbalancing the heavy shell to achieve neutral buoyancy, the large volume of the turtle's lungs is beneficial for another reason. The lungs are the turtle's major storage site for oxygen during diving. When submerged underwater, the turtle no longer has access to its normal supply of oxygen and must rely on oxygen that is already present in the body. Most of this oxygen is in its lungs, with much of the balance bound to the hemoglobin of its blood. A simple calculation can show that because of the large size of its lungs and its low rate of metabolism, a turtle can remain underwater for at least an hour during summer temperatures before its contained oxygen is depleted and it must rise to the surface for a refill. By slowing down its metabolism, the turtle can stay submerged even longer and still have an adequate reserve of oxygen; that is, it can stay within its aerobic dive limit. Of course the turtle can stay underwater even after all of the oxygen is gone by switching to anaerobic metabolism, which does not require oxygen, but that is the subject of another discussion (see Chapter 7).

Buoyancy Control in Marine Turtles

The dual function of the lungs both as a buoyancy organ and for oxygen storage has some important implications for marine turtles, as they typically dive to much greater depths than do freshwater turtles. Lung gas is compressible, so as an animal dives deeper, the (hydrostatic) pressure of the water bearing down on it increases, thus lung volume, and therefore its buoyant effect, decreases. As informed recreational divers know, hydrostatic pressure increases by one atmosphere for every 10 m of water. This means that 10 m down, the total pressure is two atmospheres, the true atmospheric pressure at the surface plus the additional atmosphere due to the hydrostatic pressure produced by the 10-m column of water. At 20 m, the total pressure is three atmospheres, and so forth. Recalling Boyle's law, a gas volume of 6 L at the surface compresses to 3 L at 10 m and to 2 L at 20 m, assuming that the gas temperature remains unchanged.

Recent studies of marine turtles, including green turtles and loggerhead turtles, suggest that they frequently dive to a particular depth and remain at that depth for the duration of the dive. The chosen depth may be a site for resting or a site for foraging. Furthermore, the turtles apparently anticipate the targeted depth prior to diving, and thus they fill their lungs with a volume of air that when compressed at the chosen depth will confer neutral buoyancy. If the turtle makes shallow dives, then surface lung volume need not be too large, but if dives are deeper, then lung inflation at the surface must be correspondingly larger. Consequently, like freshwater turtles, marine turtles are able to adjust their lung volume over a wide range in the interests of buoyancy control, although it is worth noting again that they do this without the benefit of cloacal bursae. If experimenters attach lead weights to the marine turtle's shell, then lung inflation is greater in preparation for a dive to a particular depth than it is in turtles without added weights. This is now a situation where the buoyancy and oxygen storage functions of the lungs interact.

Studies of this interaction were conducted by Hays et al. (2004) on green turtles *(Chelonia mydas)* captured on the beach at Ascension Island in the Atlantic Ocean and instrumented with time/depth recorders. A subset of these turtles was also fitted with lead weights to increase the specific gravity. Female green turtles normally lay several clutches of eggs each season, so the investigators were able to recover the recorders when the turtles returned to the beach after their inter-nesting period at sea. The investigators' analysis focused on the dives during which turtles remained at a particular depth throughout the dive, a pattern they called U-dives. Hays and his colleagues were able to glean some important information regarding the turtles' diving behavior. First, the turtles' deeper dives tended to be longer ones, even accounting for the travel time to and from depth. Second, lead-weighted turtles that dived to a certain depth remained there longer than did nonweighted turtles at the same depth. Both of these observations can be interpreted on the basis of lung volume and lung oxygen stores. In order to dive deeper, the investigators believe that the turtles filled their lungs more fully so that neutral buoyancy could be achieved at the greater depth. The larger lung volume, however, supplied more oxygen to the turtles, enabling them to remain submerged longer before oxygen was depleted. Similarly, the weighted turtles had to inflate their lungs more than the nonweighted turtles to achieve neutral buoyancy at the same depth, and this greater volume and oxygen extended the dive time of the weighted turtles. As Hays et al. acknowledge, these conclusions are inferential because they are assuming (1) that turtles are at neutral buoyancy at the chosen depth, and (2) that turtles have a consistent oxygen-depletion rate and end point to their dives. A statistical analysis of their data suggests that other factors are certainly involved in determining dive time, but that oxygen stores are a major consideration.

A more complicated result came from a laboratory-based study of diving in juvenile loggerhead sea turtles (Hochscheid et al., 2003). These investigators reported that turtles fitted with lead weights did dive for longer periods, but they said that this outcome could not be explained by increased lung volume. They observed little evidence of buoyancy control in these animals.

The green turtles that nest on Ascension Island travel there from far to the west off the coast of Brazil. (Their uncanny ability to locate this tiny island in the expanse of ocean will be discussed in Chapter 5.) The eggs they deposit in nests excavated on the sandy beach, if spared by egg-loving predators, incubate for some two months before the hatchlings emerge, claw their way to the sand's surface, and scurry down the beach into the surf. Like other hatchling turtles, both marine and freshwater, they have soft shells with little mineralization. In the water, they float at the surface and have little ability to control their buoyancy.

A study by Milsom (1975) on captive Atlantic loggerhead turtles revealed that buoyancy control develops slowly and is not fully effective until the turtles are nearly a year old. The ability to actively dive for food, however, develops more rapidly. As the hatchlings mature, their lung

27

volume increases parallel to the mineralization and density of their shells, until after eleven months their lung volume, as a percentage of body volume, has nearly doubled from its four-month size, and their ability to regulate their buoyancy is complete. Positive buoyancy in newborn turtles may be advantageous, however, because it keeps them near the surface where favored food sources (plankton and algae) can be located, plus it may help them passively follow the ocean currents that transport them on their proper migration path. Like the adult freshwater and marine turtles, a close connection exists between the shell, lung volume, and diving pattern. The result is an energetically efficient animal that is well adapted to an aquatic existence.

Lung Volume Changes during Diving

Based on the foregoing information, it is clear that when a turtle changes its depth during a dive, its lung volume and thus the buoyant effect of its lung volume will change. However, lung volume is not affected only by hydrostatic pressure. As the turtle utilizes its lung oxygen during a dive, its lungs will decrease in volume, and this will also affect its buoyancy. The loss of oxygen is partially counterbalanced by carbon dioxide entering the lungs, but during a breath hold most of the metabolically produced carbon dioxide is stored within the body fluids. Because most of the gas in the turtle's lungs is nitrogen, the decrease in the turtle's lung volume is relatively modest during its routine dives within the aerobic dive limit. However, recall that at the beginning of the chapter, I described an experiment in which a turtle breathed pure oxygen for two hours and then was submerged. In this situation, the lungs collapsed rather rapidly. The only gases present in this turtle's lungs when submergence began were oxygen, carbon dioxide, and water vapor, and together the individual partial pressures of these gases added up to the prevailing atmospheric pressure. No nitrogen was present because it had been washed out during the period of oxygen breathing. The pressure exerted by carbon dioxide and water vapor is always in equilibrium with body fluid pressure, and together they will never constitute more than a small fraction of atmospheric pressure, approximately 10%, so as collapse proceeds due to oxygen removal and these gases become concentrated, their pressures change little because carbon dioxide diffuses out of the lungs into the blood and water vapor condenses. Eventually, when all of the oxygen is gone, having been consumed by the animal's tissues, the carbon dioxide and water vapor will also disappear, because as gases they cannot exert total atmospheric pressure, and thus the lungs will fully collapse.

However, breath holding with lungs filled with pure oxygen is not a normal situation. Nitrogen is the most plentiful gas in atmospheric air, where it accounts for close to 80% of the total gas volume. It is always the major constituent of lung gas under normal circumstances. Furthermore, as noted earlier, nitrogen is inert physiologically and does not participate directly in chemical reactions within the body. Its presence prevents the rapid collapse in lung volume that occurs when the lungs are filled with pure oxygen. Nevertheless, if the submergence lasts long enough, the lungs will eventually collapse, even though nitrogen is present. The reason is that the decrease in lung oxygen pressure during breath holding is greater than the increase in carbon dioxide pressure, and this concentrates the nitrogen molecules in the lungs and increases their total pressure. A gradient is thereby produced between the pressure of nitrogen in the lungs and the pressure of nitrogen in the surrounding water, resulting in a slow diffusion of nitrogen out of the lungs and a slow, but an inexorable, collapse of the lungs.

The phenomenon of a collapsing gas volume is well known in other biological situations. For example, a gas pocket trapped within our body will gradually shrink and disappear because the total partial pressure of the gases in the venous blood and other body fluids is less than in the gas pocket that is at atmospheric pressure. Similarly, a portion of a lung that is cut off from its airway and is no longer being ventilated will collapse for the same reason.

A particularly interesting example of this principle occurs in certain aquatic insects that dive beneath the surface with a bubble of air adhering to their body surface (Rahn and Paganelli, 1968). The bubble covers several spiracles, which are the openings on the body surface that lead into the fine, air-filled channels called tracheae. The tracheae provide a pathway for oxygen to diffuse from the surface of the insect to its cells. In many diving insects, the attached bubble is collapsible, and its size depends on the net exchange of gases with the insect and with the surrounding water. The key functional exchange processes are the diffusion of oxygen from the bubble down the tracheae to the cells and the diffusion of oxygen from the surrounding water into the bubble to replenish the lost oxygen. The bubble provides a local gas phase that facilitates the transfer of oxygen dissolved in the surrounding water into the air-filled tracheae. Because the oxygen pressure in the bubble must fall below the water oxygen pressure to enable diffusion of oxygen into the bubble, the nitrogen pressure will rise in the bubble and exceed the water nitrogen pressure. This is the same thing that happens in a turtle's lungs during prolonged diving. Nitrogen will therefore diffuse from the bubble out into the water, causing the bubble to gradually shrink in size. Eventually it becomes so small that the insect must return to the water's surface to restore the bubble to its proper size. The curious aspect of this behavior is that a major reason the insect must return to the surface is to replenish the nitrogen, not the oxygen. It is the nitrogen that is required to sustain the bubble's volume and to enable it to function as an effective gas exchanger. If the insect dives deeper into the water, the hydrostatic pressure raises the bubble pressure and accelerates the depletion of the nitrogen and bubble collapse and necessitates more frequent returns to the surface.

A phenomenon similar to the diving insects also occurs in certain aquatic amphibians, particularly salamanders. They can obtain all of the oxygen they need directly from the water through their skin, and yet they swim to the surface periodically to breathe. In experiments in which bullfrog tadpoles and salamanders, mud puppies and sirens, were submerged in aerated water without access to the surface, it was noted that the lungs of these animals collapsed after eight to twelve hours at room temperature (Ultsch et al., 2004). Unlike the insects, these amphibians do not need to replenish their gas volume in order to continue normal aquatic oxygen consumption, because that is accomplished by their permeable skin or gills. Nor is it thought that maintenance of buoyancy is the reason for this behavior. Instead, Ultsch et al. suggested that the lungs are kept filled in case the water becomes low in oxygen (hypoxic) and lung breathing becomes necessary for obtaining oxygen. Breathing is easier if the lungs are already inflated.

Unlike salamanders, however, turtles as a rule do not remain submerged long enough at summer temperatures for lung collapse to occur because they must breathe periodically to replenish their lung and blood oxygen levels. An exception may be the Australian cloacal-breathing turtle, *Rheodytes,* referred to earlier. It is possible that because its aquatic gas exchange is so effective that its periodic excursions to the surface may at times be to keep its lungs inflated rather than to obtain oxygen, although this is just conjecture.

During the winter, however, when freshwater turtles can be trapped under the ice for many weeks, conditions are just right for lung collapse. Nitrogen pressure in the water can be no higher than in ambient air, but nitrogen pressure in the lungs will be higher because oxygen in the lungs is low. Thus nitrogen should slowly leak out of the lungs, leading eventually to lung collapse for the same reasons described earlier. Because the turtle's access to the surface is blocked by ice, the turtle cannot replenish the gas, thus its lungs will remain collapsed until the ice melts, and in very cold climates this may not occur for months. Some years ago, Gordon Ultsch's and my observations confirmed that this actually does occur (Ultsch and Jackson, 1982). We simulated this natural overwintering experience of northern painted turtles by submerging them in cold water (3°C) for up to five months without permitting them access to the surface. The water was bubbled with air, thereby keeping the gas pressure in the water the same as in the atmosphere. At the end of the submergence, we dissected some of these animals and found that their lungs were indeed collapsed.

In contrast, turtles that had been submerged in water bubbled with pure nitrogen gas had inflated lungs. The lungs did not collapse in this situation because nitrogen pressure in the water was abnormally high, as high as or higher than in the lungs, so nitrogen did not diffuse out of the lungs. But a nitrogen pressure this high does not occur naturally, since the pressure in water will never exceed atmospheric nitrogen pressure, even if the water loses all of its oxygen. Consequently, the lungs of turtles holding their breath in ice-covered ponds for many weeks should collapse whether the water has oxygen or not.

As an aside, it is worth noting that these overwintering turtles with collapsed lungs are now in the same state as the turtle described earlier, the one that was submerged in the laboratory with O_2-filled lungs. Recall that the cloacal bursae of the latter turtle filled with a volume of water that equaled the loss in lung volume. We now see that this is not such an abnormal situation for a turtle after all. The same compensation must occur every year as the lungs of a turtle collapse while it overwinters under the ice. Ultsch's and my observation, in our study of cold-submerged turtles, bears this out. We found, to our surprise, that the body weight of turtles that were submerged in aerated water (the turtles with collapsed lungs) increased, on average, by 11.9% of the turtles' original weight, an observation we could not explain at the time (Ultsch and Jackson, 1982). Perhaps not coincidentally, 11.9% is very close to the normal lung volume of the closely related red-eared slider. The weight gain we observed, therefore, was quite likely due to water uptake into the cloacal bursae to offset the loss of lung volume.

Let us return to the overwintering turtles. When at last the ice melts and the way to the surface is once again open, the turtle can now breathe, but it will face some challenges in doing so. First, it will be weak from its long period of inactivity and will likely experience some energy problems due to lack of oxygen (discussed more fully in Chapter 6). Second, the turtle will be heavy in the water with no lung gas to provide buoyancy, not unlike the turtle described at the beginning of the chapter, the one that had the lead weight taped to its shell. Some effort will be required by the turtle to swim to the surface of the water, an effort that may be hard to muster after a winter of cold submergence. Moreover, when at last the turtle does reach the surface, it will have to inflate its lungs from the collapsed state, a situation not unlike the first breath of a human infant at birth. As in the neonate, this will require a burst of energy. The effort required for the dramatic first breath by the turtle is reduced, as it is for a human baby, due to the presence of a chemical agent called surfactant, which prevents the collapsed lung surfaces from adhering to each other and makes the lungs easier to inflate (Daniels and Orgeig, 2003). If the turtle can successfully meet all of these challenges, then a few breaths at the surface can permit it to reinflate its lungs, expel the excess water from its cloacal bursae, restore normal oxygen to its system, reestablish its normal buoyancy, and prepare it for another season of activity.

3

THE BREATHING TURTLE

One would assume that breathing must be difficult and challenging for an animal encased in a rigid shell. And, indeed, it would be if the poor creature were completely enclosed. But a turtle has soft, flexible pockets of skin where its limbs emerge. These pockets move in and out as the turtle expires and inspires. As we will see, the turtle in its shell can not only breathe but can do so at quite a modest metabolic cost, despite its rigid enclosure. Because of its shell, the breathing mechanism of the turtle is unusual. Unlike other reptiles that employ trunk muscles associated with the ribs for breathing, the turtle must recruit other muscles for the purpose, since its ribs are fused to the inside of its shell. The muscles that turtles, in general, employ are abdominal ones, which either compress the lungs, producing expiration, or expand the posterior limb pocket, producing inspiration (Landberg et al., 2003). Despite the proximity of these muscles to the hind limbs, breathing, at least in some turtles, is relatively independent of limb movements.

The respiratory system of a turtle must adapt to the varied circumstances of its existence. The turtle may be at rest, or it may be active; it may be on land with continual access to air, or it may be in water, where its breathing is interrupted by extended stays beneath the surface; and, because a turtle is cold-blooded, its body temperature may range from near freezing during a cold winter to well over 30°C during a hot summer. Each of these circumstances will be discussed, but temperature change is a good place to begin our discussion of breathing because temperature affects so many aspects of a turtle's physiology.

How Temperature Affects Breathing

My brief scientific career before I began studying turtles centered on temperature physiology. My PhD thesis with Professor H. T. (Ted) Hammel at the University of Pennsylvania was a study of temperature regulation in exercising dogs, and my initial project during my postdoctoral stay with Professor Knut Schmidt-Nielsen at Duke University concerned countercurrent heat exchange in the nasal passages of kangaroo rats. Both dogs and kangaroo rats are of course mammals that regulate their body temperatures at a relatively constant value. Turtles, on the other hand, are cold-blooded animals whose body temperature approximates the temperature of their immediate surroundings. When I decided to study breathing in turtles, my background in temperature physiology influenced my decision to begin by testing the effect of temperature, although as I will reveal shortly I had another reason for this choice.

My study was a straightforward one. My goal was to learn how much a turtle, in this case a red-eared slider, breathed per minute at 10°C and 30°C. I measured the metabolic rate (that is, oxygen consumption) and breathing volumes of turtles at these temperatures. As expected from many previous studies, I found that the rate at which the turtles consumed O_2 was about four times greater at 30°C than at 10°C. An increase of this magnitude in metabolic rate typically occurs when the body temperature of a cold-blooded animal increases by 20°C. The breathing volumes, however, were quite surprising, particularly for someone who mainly thinks about mammalian or human physiology. I observed that the volume of air the turtles breathed at the two temperatures was nearly the same (Jackson, 1971a). Why is this surprising? Well, consider what happens when you go for an easy jog, an activity that might increase your oxygen consumption fourfold. To supply this extra O_2, you increase your breathing. If you doubt that this happens, then try resisting the urge to breathe more the next time you exercise and see how you feel. Numerous studies have found that the increase in breathing during activity is not haphazard but closely matched to the rate of metabolism. For example, if the metabolic rate increases fourfold, then the volume of air breathed per minute also increases approximately fourfold. Matching breathing to oxygen consumption serves an important function in our bodies; it stabilizes the composition of our lung gas, specifically the concentrations of oxygen and carbon dioxide. Holding lung gas composition constant, (and, indirectly, blood gas composition) contributes to the maintenance of a relatively constant environment for our cells, called physiological homeostasis.

Given the homeostatic principle I have just described, the response of the turtles to an elevated metabolic rate demonstrated that these ancient animals do not always obey the "rules" of human physiology. Indeed, from the perspective of human physiology, the turtle at 30°C compared to the turtle at 10°C was underbreathing, that is, hypoventilating. Hypoventilation can be a serious clinical condition for a human patient because not enough fresh air is being breathed in to keep pace with the consumption of oxygen (O_2) and production of carbon dioxide (CO_2). Consequently, lung O_2 is lower than it should be and lung CO_2 is higher than it should be, and because blood passing through the lungs equilibrates with the lung gas, the same departures from the normal state exist in the arterial blood as well. By convention, lung and blood gases are expressed as partial pressures (the portion of the total gas pressure exerted by a particular gas), so hypoventilation leads to low O_2 pressure (low PO_2) and high CO_2 pressure (high PCO_2). To compound the problem even more, high blood PCO_2 produces acidosis, a low blood pH, a condition that if sufficiently severe can depress brain function and produce a comatose state.

So what was the problem with the turtles in my study? Did they need medical attention? Were they not maintaining normal homeostasis? No, the turtles were not sick—their breathing was normal at each temperature; and yes, the turtles were defending their homeostasis. However, the physiological state of the turtles in my experiment and your physiological state when you exercise are fundamentally different. Your metabolic rate increased because of exercise at essentially a constant body temperature, whereas the turtle was at rest throughout, and its metabolic rate increased at 30°C because of a higher body temperature.

My experiment simulated a common experience for a freshwater turtle. Imagine a mild day in early spring. A turtle is floating near the surface of a pond. The temperature of the water is still cool, at 10°C. Because the turtle is a cold-blooded animal, its body temperature is also approximately 10°C. The turtle climbs out of the water onto a floating log and sits quietly basking in the warm sun. Slowly its body temperature rises, perhaps eventually reaching 30°C. Outwardly, it looks calm and its activity level has not changed, but inside of its body the higher temperature has caused the rates of chemical reactions to accelerate, causing the turtle's overall rate of metabolism to increase.

In addition, other important chemical changes have occurred that have profound effects on the body's acid-base state, indicated most commonly by the pH of the blood. The chemical effects of temperature include changes in the solubilities of dissolved substances and in the equilibrium states of many chemical reactions. The net effect of these chemical changes has produced a fall in the blood pH of the basking turtle as well as a rise in its blood PCO_2. These new blood values represent the correct homeostatic condition for the higher temperature. The breathing response I observed was the physiological mechanism the basking turtle uses to keep its blood pH and PCO_2 at their new and proper values.

In the interests of full disclosure, I must now admit that the results from my experiment with the turtle were actually not surprising to me. Instead, I would characterize my response more as delight or even relief, because the results confirmed what I predicted would happen. I was inspired to perform this study after reading a paper (Howell et al., 1970) that reported the values of blood pH and blood PCO_2 in a variety of cold-blooded animals, including the snapping turtle, at various body temperatures. As in the earlier hypothetical example of the basking turtle, the average pH of the snapping turtles in this study fell consistently and PCO_2 rose consistently when the body temperatures of the turtles were raised, stepwise, from 5°C to 35°C. Being a reasonably well-informed respiratory physiologist, I realized that these values, specifically the large increase in blood PCO_2 at 35°C, were clearly indicative of "hypoventilation" at that temperature, but to my knowledge no one had directly tested how the breathing of a cold-blooded animal is affected by temperature.

At the time, I was completing my studies on buoyancy control, described in Chapter 2, and those studies suggested a simple way I could measure the breathing of a turtle that would enable me to test how temperature affects the breathing. I suspended a turtle into water from a strain gauge, a device that measures force. In this application, the force I measured was the turtle's underwater weight. I added a lead weight to the turtle so that it was always heavier than water, and then I wrapped tape around the turtle so it was unable to swim away. The tape did not disturb its breathing. I positioned the turtle so that it could easily elevate its head above the water's surface and breathe from an inverted box through which air flowed at a measured rate. By analyzing the composition of the air leaving the box, I was able to measure metabolic O_2 consumption. When the turtle breathed, its underwater weight, or buoyancy, changed, and this was detected by the strain gauge and recorded on a polygraph to provide a measure of the volume breathed. By regulating the temperature of the water, I could set the turtle's body temperature where I wanted it. Using this method, I was able to confirm that at a high temperature turtles do indeed "hypoventilate." But I must hasten to emphasize that this is not hypoventilation in the human medical sense; it is normal breathing for a warm turtle.

Blood pH and Body Temperature

The study by Howell and her colleagues (1970) was not the first description of the effect of temperature on the acid-base properties of cold-blooded animals. Earlier studies dated back to the 1920s; but in 1962, Eugene Robin published the first comprehensive study on the subject, and his was the first paper to have a major impact on the field of comparative animal physiology. Robin studied the red-eared slider, a common species that has been the subject of many physiological studies, including my own. He sampled blood from turtles that had been held for a period of time (that is, they had been acclimated) at two different temperatures, and he then observed changes in blood pH and PCO_2, similar to those in the study of Howell et al. (1970). In addition, Robin collected blood from turtles and equilibrated the blood at different temperatures in test tubes outside of the body. When he measured pH and PCO_2 in these test-tube samples, he found that the pattern of change was almost identical to what he had observed on blood taken directly from the acclimated turtles. This demonstrated a fundamental principle: the intact turtle regulates its blood acid-base state to closely match the temperature-dependent chemical changes that occur in a blood sample equilibrated to different temperatures outside of the body.

Consider again the turtle sunning itself in the pond. Imagine that when it crawls onto the log, its body temperature instantly rises from 10°C to 30°C. The chemistry of its blood shifts to a new thermal state: blood pH spontaneously falls, and blood PCO_2 rises. This mimics what Robin observed in a sealed test tube, but now it has happened in a living, breathing turtle that produces CO_2 in its cells and loses CO_2 by breathing. How much the turtle breathes relative to its metabolic CO_2 production will determine its steady-state blood pH and PCO_2. Suppose the turtle acted like we do when we exercise at a constant temperature and increased its breathing to match the increase in metabolic CO_2 production. Its blood PCO_2 would revert to the low 10°C PCO_2, and its pH would rise to its higher 10°C value. These would be the wrong values for 30°C, and homeostasis would be severely disturbed. However, by keeping its breathing nearly the same or increasing it only slightly, the turtle defended the new lower pH and the new higher PCO_2, the appropriate values for this temperature. "Hypoventilation" has saved the day. In reality, of course, the turtle's body temperature changes slowly, and the direct chemical effects of temperature and the turtle's physiological breathing adjustments occur simultaneously.

It may be obvious by now that an invariant blood pH and PCO_2 are not characteristics of normal homeostasis at different temperatures in turtles. We are conditioned to these values staying constant because that is how it works in our bodies, which are always close to a constant temperature of 37°C. In fact, a blood pH of 7.4, the normal human blood pH, was long considered the gold standard, no matter what the temperature. We now know that the sacred pH of 7.4 is only appropriate at a temperature near 37°C. A turtle, on the other hand, functions over a wide range of body temperatures, and its homeostasis, including its blood pH and PCO_2, adjusts accordingly.

A fundamental question remains, however: What are turtles and other animals that exhibit this behavior regulating homeostatically when body temperature changes if it is not the absolute value of blood pH? An early suggestion was that the animals maintain a constant difference between blood pH and the neutral pH of water, what was called "constant relative alkalinity" by Hermann Rahn of the State University of New York at Buffalo (Rahn et al., 1975). Neutral pH (neutrality) is usually thought to be at a pH of 7.0, but neutrality actually changes with temperature, from approximately 6.8 at our body temperature of 37°C to 7.45 at 1°C. As shown in Figure 3.1, this temperature-dependent change in neutral pH parallels the change in the blood pH of cold-blooded animals.

The current consensus among physiologists, however, is that the fundamental homeostatic property that animals are regulating concerns the structure and function of proteins in the body. Proteins, which are composed of long chains of amino acids, have ionized groups, or electrical charges, within their structure that help determine the three-dimensional shape of the protein. This three-dimensional shape, in turn, is critical to the normal functioning of the protein, and proteins of course perform a multitude of functions within a living organism. If the pH of the solution in which the protein resides changes, then the electrical charges are altered, thereby disturbing both the structure and function of the protein.

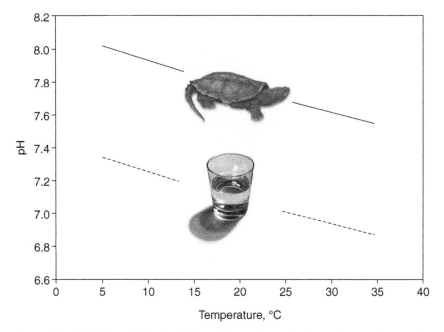

Figure 3.1 The blood pH of cold-blooded vertebrates such as the snapping turtle changes with temperature in parallel to the neutral pH of water. Illustration of turtle courtesy of Toronto Zoo Adopt-a-Pond Programme, www.torontozoo.ca/adoptapond. Artwork produced by Wallace Edwards.

This is a major reason careful regulation of pH within an animal's body fluids is so important. The most familiar fluid is blood, and it is the pH of blood that I have been describing, but this crucial regulatory process applies as well to the pH of the fluid within body cells, the so-called intracellular pH. At constant temperature, such as in our own bodies, a narrow pH range, in the vicinity of 7.4 for our blood or somewhat lower within our cells, satisfies this homeostatic requirement for protein function. In a cold-blooded animal such as a turtle, however, the pH values necessary to maintain the integrity of protein structure and function change with temperature. For a turtle at 37°C, for example, the correct blood pH is close to 7.4, just as it is for us. However, at a usual summer temperature of 20°C, the pH is about 7.7, and in a cold winter pond, the turtle's blood pH is about 8.0. At each of these temperatures, even though the pH values are very different, the ionization state of the proteins is about the same, and the structure and function of these proteins are relatively unchanged. Readers interested in a more detailed account of the history of this idea and the physical chemistry underlying it can consult a review by Reeves (1977).

This description of temperature-dependent acid-base regulation is often called the "alphastat hypothesis" (Reeves, 1972), because "alpha" was used as the symbol for protein ionization, and "stat" means unchanged or stationary. Because of the innumerable jobs that proteins perform in living cells, optimizing their functional characteristics in the face of large changes in temperature represents a crucial aspect of homeostasis (Somero, 1986). The regulation of pH in response to body temperature change as observed in turtles and in many other cold-blooded animals is thought by many to satisfy this homeostatic requirement.

Later studies have generally supported the alphastat hypothesis (Ultsch and Jackson, 1996), although in physiology, as in life generally, things are never quite as simple as they first appear. For example, in most subsequent studies of cold-blooded animals, including other turtles, breathing actually does increase when temperature rises, but almost never as much as metabolic rate (Jackson, 1982). This still represents a relative hypoventilation and is the response that is required to sustain a decrease in blood pH and an increase in blood PCO_2 as temperatures rise. A more important challenge to the alphastat hypothesis, and the basis for considerable controversy about the hypothesis, is that blood pH and particularly intracellular pH do not always follow the ideal relationship described by Reeves and his colleagues (Heisler, 1986), and in some animals, such as monitor lizards and hibernating mammals, blood pH is relatively independent of temperature. Furthermore, the constancy of protein charge within cells is difficult to verify (Cameron, 1989), and some animals do not maintain constant levels of whole body CO_2, as alphastat theory predicts, when body temperature changes (Stinner et al., 1994) . As is the case with most aspects of biology, our understanding of these complex events is still incomplete, although alphastat regulation remains the generally accepted theory.

How Exercise Affects Breathing

Although the response to temperature change produced a surprising breathing response by the turtles, they behave much like we do if their metabolic rate increases as the result of increased activity at a constant temperature. Like us, they tend to keep their blood pH at a constant value under these conditions. Unlike us, on the other hand, breathing and exercise may in some cases come into conflict in turtles and other reptiles.

As described more fully in Chapter 4, I took several research trips to Tortuguero Beach, Costa Rica, with my colleague, Henry Prange, to study the respiratory physiology and energetics of nesting green sea turtles, *Chelonia mydas*. As part of these studies, we measured the metabolic rate and breathing volume of turtles at rest and during normal beach activities, including walking and nesting. When we averaged these measurements over a period of several minutes, we learned that green turtles when active at a constant temperature increased their breathing roughly in proportion to the increase in metabolic rate, just as we do. However, we observed a curious pattern of breathing in these turtles; they only breathed during pauses in their activity. While walking, probably better described as hauling, a green turtle typically makes several laborious forward movements, then stops completely and takes several large breaths, each breath beginning with a strong exhalation following by an inhalation. This alternating pattern of activity and breathing was also the pattern we observed when a turtle excavated its nest. We reported this phenomenon as an incidental observation (Jackson and Prange, 1979), and although we did not attempt to study its basis, we postulated that one possible explanation was that some of the green turtle's muscles contribute to both breathing and locomotion, thus the two activities could not occur simultaneously.

At about the same time, an undergraduate student, David Kraus, set up in my lab at Brown University a swimming tank, a flume, with the intention of studying swimming metabolism in juvenile green turtles. A flume is like a treadmill for swimming. Water flows in a channel and the turtle must swim to maintain its position in the water, just as a person on a treadmill must run to remain in the same location. Because our objective was to measure the swimming turtle's O_2 consumption, we covered the surface with a sheet of stiff plastic, except for a single location where the turtle could raise its head above the water to breathe. Training a turtle to swim in the flume with the surface open was not difficult, but a problem arose when we installed the plastic sheet with the breathing hole. We discovered that the turtle swam just beneath the surface, but when it decided to breathe, it stopped swimming and elevated its head. This would work fine out in the ocean. The turtle would continue to coast forward as it raised its head and took a breath, and then it would lower its head and continue swimming. However, in our flume, the current swept the coasting turtle toward the back of the flume behind the breathing hole, and thus the turtle was unable to breathe. Before we were able to figure out a solution to this dilemma, the U.S. government banned all imports of marine turtles into the country, and we had to discontinue the project.

In retrospect, we had not read with sufficient care a paper published by my colleague, Henry Prange (1976), who had studied, several years earlier, young green turtles swimming in a flume. He provided a long breathing chamber so that a turtle still had access to the surface even as it floated back during its pause to breathe. In a subsequent study, Butler et al. (1984) successfully accomplished what David Kraus and I were attempting, plus more, when they measured the metabolic rate of swimming green turtles in a flume. Like Prange, they also used a long breathing chamber, but even so they found that when water flow reached a critical velocity, the turtles were swept to the back of the test chamber and were unable to breathe.

Why should a green turtle not be able to breathe and exercise at the same time? Certainly we and other mammals can breathe as we run, and birds can breathe as they fly. The reason is that we use different muscles for the two functions. Mammals evolved a diaphragm that is committed to breathing and not involved with running. Indeed, locomotion, whether running or flying, enhances breathing in many animals. Perhaps performing the two functions simultaneously is only a problem for lower vertebrates, such as reptiles.

This possibility led David Carrier, then at the University of Michigan, to study breathing in four species of lizards during treadmill exercise (Carrier, 1987). His consistent observation in each species was that ventilation increased at a slow speed but then progressively decreased as the speed of running increased further. This is precisely opposite to what happens when we exercise and is counter to our expectation that when an animal needs more O_2, more breathing is recruited to supply the O_2. In subsequent studies, Carrier was able to show that a conflict exists between the use of the lizard's trunk muscles for both breathing and running. Breathing requires that these muscles contract synchronously on both sides of the body to generate the pressures within the body that cause air to move in and out of the lungs. Running, on the other hand, requires that some of the same muscles contract in an alternating fashion, first one side and then the other, to twist and support the body wall. A lizard cannot effectively perform both actions simultaneously. Carrier noted that the largest breathing volumes actually occurred after the treadmill stopped and the lizard returned to the resting state. Now no conflict existed, and the muscles could be fully committed to breathing.

Lizards in nature typically engage in burst-type activity, running at a high speed and then stopping to rest. Much of their energy during these bursts is supplied by anaerobic metabolism (which does not require oxygen), the same source of energy that supports human athletes during intense efforts, such as the 100-m dash.

However, some lizards, such as monitor lizards, are more aerobic and can sustain activity for a longer period of time, in the way a long-distance runner can. Two of the species studied by Carrier were monitors, and it was unclear how sustained aerobic activity could be possible, given the constraint on breathing that he described. This paradox was resolved when a study by Owerkowicz et al. (1999) found that the savannah monitor, *Varanus exanthematicus,* whose normal habitat is sub-Saharan Africa, supplements its breathing during running by pumping air into its lungs from its throat, a breathing mechanism called gular pumping. The monitor lizard, like the other lizards studied by Carrier, has the same constraint on aspiration breathing by the trunk muscles during running, but it circumvents the problem and sustains its breathing by gular pumping.

These intriguing observations have evolutionary significance. They suggest that the ancestors of amniotes, the forerunners of reptiles, birds, and mammals, had the same conflict between breathing and running as described by Carrier, but that various lineages evolved ways to deal with the problem. Mammals and birds evolved breathing mechanisms that were largely uncoupled from locomotion, and monitor lizards utilized positive pressure gular pumping that resembles the mechanism, called buccal pumping, used by amphibians and some air-breathing fish. But what about turtles? Because of their rigid shells and the incorporation of their ribs into the shells, the same constraint on breathing seen in the lizards could not be an issue for turtles.

Prange and I made our observations on a marine turtle, but what about a freshwater turtle? Does it also stop breathing when it exercises? Landberg et al. (2003, 2009), in their work at the University of Massachusetts at Amherst, addressed this question. They trained box turtles and red-eared slider turtles to walk on a treadmill while wearing a helmet connected to a device called a pneumotach, which provided a continuous measurement of breathing volumes. Their results demonstrated clearly that these turtles are able to walk and breathe at the same time. Unlike the lizards, they do not have a conflict with these two functions. It has long been thought that breathing in turtles is associated with movement of the limbs in and out of the body cavity, so one possibility for the results of Landberg et al. was that walking and breathing are coupled and synchronized. However, careful measurements showed that this was not the case; rather, the two are independent of each other. The authors concluded that the freshwater turtles use distinct muscles for walking and for breathing, and that their actions neither conflict nor interact. A graduate student. Daniel Warren, who worked in my lab, studied painted turtles swimming in a flume similar to the one used previously for the green turtles, and his observations, including a video of a swimming turtle, clearly showed that a painted turtle can swim and breathe at the same time.

This still leaves unanswered the question concerning the green turtle and other marine turtles. Are they really not able to swim (or walk) and breathe simultaneously, and if not, then why are they different from freshwater turtles that do not have the conflict? Further studies are needed to resolve this. Tortuguero, anyone?

Intermittent Breathing and Respiratory Control

Although freshwater turtles such as red-eared sliders can swim and breathe simultaneously, their normal activity pattern often resembles that of the green turtle. They swim around underwater and periodically come to the surface, where they float quietly and take several breaths before returning again to their underwater pursuits. This normal behavior dictates that their pattern of breathing is quite different from the way we breathe. We generally breathe continuously, with one breath following another in a regular rhythm. However, this is not the case in freshwater turtles. When I made the measurement of a turtle's breathing, described earlier, the turtle was free to breathe at will, yet it continued to breathe intermittently, taking a series of breaths and then holding its breath for several minutes. A representative breathing record illustrating this pattern is shown in Figure 3.2.

Intermittent breathing, which has been observed by everyone who has studied freshwater turtles, persists whether the turtles are in water or on

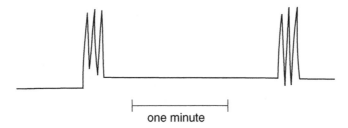

one minute

Figure 3.2 Intermittent breathing record of a red-eared slider turtle. Each breathing episode depicted begins with an expiration and has three breaths. A period of nonbreathing (apnea) follows each episode.

land. We now have evidence that this pattern is intrinsic to its neural control system. Douse and Mitchell (1990) performed a study in which they removed the brain from a turtle, isolated the brain stem and upper spinal cord, and placed this preparation in a chamber with a solution that resembles the fluid that normally bathes the brain in order to keep the brain tissue alive. They then located particular nerves leaving the brain and spinal cord and recorded the spontaneous electrical activity, or action potentials, in these nerves. Their results revealed a burstlike pattern of neural activity that closely resembled the intermittent breathing pattern of intact turtles. They also recorded from some of these same nerves in an intact, anesthetized turtle and observed similar bursts of activity that were synchronized with the turtle's episodic breathing. Like humans and other vertebrates, turtles possess a complex network of nerve cells in the brain stem that generates a respiratory rhythm. But whereas our rhythm is regular, with evenly spaced breaths, the turtle's is intermittent. The 1990 study by Douse and Mitchell suggests that this pattern is hardwired into a turtle's nervous system.

William Milsom of the University of British Columbia carried out detailed studies of intermittent breathing in a variety of animals (Milsom, 1991) and found that this pattern is common to many reptiles, amphibians, and air-breathing fish. It also occurs in otherwise continuously breathing mammals when they hibernate at a low body temperature. Milsom dissected the intermittent pattern into a ventilatory period and a nonventilatory period. He and others have recognized that the turtle's respiratory pattern, even though, as noted earlier, it is probably hardwired, is modulated by input to the brain so that breathing is appropriately matched to the circumstances of the animal, such as changes in activity or body temperature. What are the signals that terminate breathing at the end of the ventilatory period and initiate breathing at the end of the nonventilatory period? The best guess is that the animal's respiratory control center responds to changes in blood O_2 and CO_2, although this has been difficult to demonstrate conclusively. Because the act of breathing raises blood oxygen pressure (PO_2) and lowers blood CO_2 pressure (PCO_2), a reasonable hypothesis is that breathing ceases when PO_2 is high enough and/or when PCO_2 is low enough during the ventilatory period. During the nonventilatory period, PO_2 gradually falls and PCO_2 rises as the turtle's cells consume O_2 and produce CO_2, and breathing may resume when these values reach critical thresholds. One piece of evidence supporting this hypothesis is that turtles in water that were free to breathe at will invariably returned to the surface to breathe before blood PO_2 fell below a critical value of about 20 mmHg (Ackerman and White, 1979). Additional support for the threshold hypothesis is that if an experimenter holds a turtle's lung and blood PO_2 high and PCO_2 low by artificially ventilating the lungs, then the animal will make no effort to breathe (Kinney and White, 1977). Breathing is also turned off naturally when a turtle at low temperature can obtain enough O_2 directly from the water and has no need to breathe air.

Conscious Control of Breathing

Let's think about our own breathing again for a moment. Threshold changes in blood gases are also thought to stimulate our breathing, and this occurs without any conscious effort on our part. But we also breathe because of input to the respiratory control center from higher brain centers in the cerebral cortex. Clear evidence for this is that whenever we wish to do so, we can voluntarily increase or decrease our breathing. Control from higher centers in the brain does not necessarily require conscious awareness or intention, however. I saw evidence of this when I once asked students in a physiology laboratory to breathe deeply and rapidly for a minute or so, a maneuver called hyperventilation. The effect of hyperventilation, the opposite of hypoventilation, is to flush out the lungs with fresh air and thereby raise lung PO_2 and depress lung PCO_2. Blood flowing through the lungs equilibrates with the lung gas so arterial blood PO_2 and PCO_2 change in the same way. If threshold stimulation from these gases were the only reason we breathe, then the expectation would be that following hyperventilation the students would simply stop breathing until these gases returned to their effective levels. But this was not what happened. Instead, the students, who were not told what to expect, blissfully continued to breathe without missing a beat. No doubt some decrease in breathing occurred, and over time their blood gases returned to normal, but the act of breathing was clearly not fully dependent on these chemical factors.

Evidence for this dual drive to breathe in humans is also found in a curious clinical condition known as "Ondine's Curse." The name is derived from a myth involving a water nymph called Ondine. The story is worth recounting. The beautiful Ondine fell in love with a handsome mortal, married him, and bore his child, even though by doing so she lost the immortality that was a perquisite of her water nymph status. She made this crucial decision because he vowed "to love her with every waking breath," and she was so smitten by this pledge of devotion from her lover that she submitted to him, despite the enormous sacrifice it required. As time went by, her loss of immortality caused her to age and to lose her youthful beauty. Her husband's passion for her waned, and his eyes began to wander. One day Ondine caught the scoundrel betraying her with another woman. In her anguish and fury she used his own oath, that "he would love her with every waking breath," to curse him by declaring that he would only breathe as long as he was awake. If ever he fell asleep, he would stop breathing and die. And so it was.

A patient who suffers Ondine's Curse, fortunately a rare condition, tends to stop breathing when asleep and must be connected to a ventilator to avoid the fate of Ondine's unfaithful lover. The underlying basis of the illness is an absence or severe reduction in the chemical drive to breathe and a much greater dependence on the conscious drive to breathe from higher centers. Unfortunately for these patients, the conscious mechanism stops operating during sleep when continued normal breathing depends on a functional chemical control system. In other words, when we sleep we revert to being turtle-like, at least with respect to this aspect of our breathing. A turtle appears to rely entirely on adequate chemical stimulation to breathe, and in the absence of this stimulation, breathing stops. Even though a turtle has an intrinsic respiratory pattern generator in its brain stem, the output from this brain center can be turned off when blood gases are not at threshold levels.

Breathing Response to Low O_2 and High CO_2

A curious aspect of our breathing and that of other air-breathing animals, including turtles, is the relative insensitivity to decreases in inspired

O_2. This may be surprising, considering the critical importance of O_2. Indeed, inspired O_2 must be well below normal before either a human or a turtle increases its breathing. As I discovered (Jackson, 1973), turtles have a curious temperature-dependence to their low O_2 response that ties in with their pH/temperature relationship. Recall from the earlier discussion that in order to maintain the correct pH and PCO_2, a turtle breathes about the same volume of air per minute at 30°C as it does at 10°C, even though its metabolic rate at 30°C is four times higher. One way to look at this, as I did earlier, is to suggest that the warm turtle is underbreathing, or "hypoventilating." But from the standpoint of O_2 uptake, it is also appropriate to say that the cold turtle is overbreathing, or "hyperventilating." At 10°C, a turtle's lung PO_2 is quite high, because the volume of air breathed is large in relation to the rate of O_2 consumption. Consequently, a turtle at this low temperature is quite indifferent to reduced O_2 in the air and does not increase its breathing until the inspired O_2 falls to 3% (normal inspired O_2 is 21%). As an aside, it is worth noting that we would rapidly succumb were we to breathe a gas with O_2 that low. At 30°C, where the turtle's lung O_2 is lower but still adequate to oxygenate the blood, the O_2 threshold is higher, and increased breathing occurs with 10% inspired O_2.

The O_2 threshold at 10°C is so low that a turtle would not need to increase its breathing at this temperature, even if it were on the summit of Mount Everest, where the available O_2 is equivalent to approximately 7% O_2 at sea level. At this altitude, a fully acclimatized human climber breathes five times as much as at sea level (West et al., 1983), driving blood PCO_2 down to 20% of its sea-level value. To my knowledge, despite this impressive resistance to the low O_2 environment on the summit, no turtle has yet attempted to climb this mountain, but it is worth recalling in this regard that the temperature is usually a bit colder than 10°C. One must conclude, therefore, that despite impressive resistance to a low-oxygen environment, climbing Mount Everest may not be a reasonable goal for a cold-blooded animal, even an exceptional one like a turtle.

Because a turtle spends so much time holding its breath during its nonventilatory periods, one might assume that its drive to breathe is rather weak compared to ours. But based on the vigor with which a turtle responds to elevated CO_2 in its inspired gas, this is not a safe assumption. Supplying an animal (or human) with gas to breathe that has a higher than normal CO_2 concentration is a common method for assessing respiratory responsiveness. In contrast to the relative insensitivity of our breathing to low O_2, we and other air breathers are quite sensitive to small changes in inspired CO_2. When we tested the red-eared slider in this way, we found that breathing increased tenfold when inspired gas, which is normally 0.04% CO_2, contained 6% CO_2 (Jackson et al., 1974) and even more with higher levels of CO_2 (Silver and Jackson, 1985). An increase in breathing of this magnitude is as great as the response of a human subject to this level of CO_2.

Why should a turtle be so responsive to high CO_2? My guess is that the combination of elevated CO_2 in the blood and low blood O_2 provide a powerful stimulus to breathe after a turtle has held its breath for a long time underwater. A vigorous breathing response during recovery can rapidly restore the blood gas values back to normal. This also suggests, however, that during a long, voluntary dive, the turtle's responsiveness to these changes in blood gases must be inhibited in some manner.

In mammals, the sensing mechanism that mediates the respiratory response to elevated CO_2 is located principally in the brain stem, where specialized nerve cells increase their firing rates in response to CO_2 and specifically to the decrease in local pH produced by elevated CO_2. As discovered by Bernard Hitzig, who at the time was a graduate student in my lab, the red-eared slider has a very similar receptive structure in its brain stem (Hitzig and Jackson, 1978). Hitzig experimentally altered the pH within the brain ventricles (fluid-filled chambers in the brain) of unanesthetized turtles that had previously been surgically prepared while anesthetized, and he observed increased breathing when the pH of this fluid, the cerebrospinal fluid (CSF), was reduced. This simulates what happens when PCO_2 is high. The receptors in the turtle's brain stem were quite sensitive to small changes in pH, similar to what had been observed during comparable experiments in a mammal, the goat (Pappenheimer et al., 1965). This observation on the turtle demonstrated that this brain mechanism for chemical regulation of breathing is an ancient one in vertebrate evolution.

Metabolic Cost of Breathing

Breathing is not free. It takes energy to contract the muscles that move air in and out of the lungs. One measure of the cost of breathing is the percent of the animal's total oxygen consumption used in the act of breathing. In humans the cost is estimated to be only 1%–2%. This means that if our whole body consumes O_2 at the rate of 300 mL per minute, then only about 3–6 mL of that O_2 are necessary to supply the muscles that move air in and out of our lungs. If you voluntarily increase your breathing, though, then the cost increases steeply and can exceed 25% of your total O_2 consumption at high rates of breathing. The cost represents the energy required to perform the work of breathing, and cost exceeds work because efficiency (work/cost) is less than 100%. The work done in our breathing is largely due to two factors: inflating the lungs and chest wall, and overcoming the resistance to gas flow in the airways. The lungs and chest wall are elastic structures, and to inflate the lungs they must be stretched, much like a spring is stretched when force is applied to it. The cost to perform these work functions depends on the efficiency of the muscles, which in humans is about 10%.

What about the cost of breathing in a turtle? As suggested at the beginning of this chapter, one might suppose that the cost would be high because of the rigid shell surrounding the turtle's body, and furthermore that the cost would be particularly high at a low temperature, when the breathing volume relative to the metabolic rate is high. Sure enough, a study that estimated cost by measuring the reduction in a turtle's O_2 consumption when its breathing was stopped by artificial ventilation concluded that the cost is quite high, reaching an estimated 35% of metabolic rate at 10°C (Kinney and White, 1977). When later investigators calculated that the physical work required for a turtle to breathe is rather low, they had to postulate, based on the high-cost estimate, that breathing efficiency was extraordinarily low in turtles (0.1%–0.25%), as much as a hundred times less efficient than in a mammal (Milsom, 1989).

Is this reasonable? When I read this, I had considerable doubts, not about the calculation of the work involved, which my own experiments had also suggested was low, but about the published estimate of the cost of breathing and the associated low calculated efficiency. As noted earlier, I had recorded breathing volumes of resting turtles up to ten times the usual value when they inspired gas with a high CO_2 concentration. In the same experiments, I also measured the rate at which O_2 was consumed by the turtles, and the rate was not significantly greater at the high rates of breathing than at the normal low rates, even when the measurements were made at 10°C. If the earlier estimate of the high cost was correct, then the metabolic rates of the turtles during CO_2 breathing should have been several times higher.

Therefore, with students Josh Singer and Paul Downey, I performed an additional study carefully repeating this measurement of O_2 consumption during CO_2 breathing, and I concluded that the cost of breathing is actually quite low in a turtle, difficult to detect at all, but probably less than 1% (Jackson et al., 1991).

Why was our estimate of the cost so much lower than Kinney and White's? I believe that the method these authors used, although reasonable in principle, was inappropriate for this purpose, because it is not possible to assume that all oxygen-consuming processes in their animals remained unchanged when they turned off the breathing. In my experiment, a large, tenfold increase in breathing should have resulted in a significant rise in metabolic rate if the cost of breathing were actually high. The observation that it did not strongly supports a low cost in this animal.

The low cost of breathing actually makes sense based on the structure of a turtle's lungs, which is rather different from ours (Perry and Duncker, 1980). Our lungs are homogeneous, which means that our lung structure is similar throughout. A single large airway, the windpipe or trachea, branches repeatedly until it finally terminates after some twenty-five to thirty branchings into 300 million or so similar-size, tiny saclike structures called alveoli (singular, alveolus), where gas exchange between gas and blood occurs. In contrast, a turtle's lungs are heterogeneous, with a large, central airway giving rise laterally to chambers that are subdivided by septa into small gas-exchange units called ediculae (singular, edicula), which are considerably larger than our alveoli. Portions of the turtle's lungs have even larger air spaces that are more easily inflated. Here is where much of the volume change occurs during breathing, and the greater ease of inflating and deflating these larger air spaces helps explain why breathing is inexpensive for a turtle. Most of the gas exchange occurs in the less distensible region with smaller air spaces that are also better supplied with blood vessels. The structure of the lungs of the red-eared slider was described quantitatively by Perry (1978).

Conclusions

As we have seen in this chapter and in Chapter 2, the lungs of turtles are versatile structures that serve as both buoyancy organs and as gas exchangers. The lung volume of a freshwater turtle such as the red-eared slider is large in comparison to other vertebrates, a necessity in order to counterbalance the sinking tendency of its heavy shell. The large volume of the lungs confers a further advantage by providing the turtle with its major reserve of oxygen to help support its metabolism through a long dive. The turtle has sophisticated control of its breathing, which enables it to function over a wide range of temperatures and to engage in vigorous activities.

4

TORTUGUERO

I vividly remember walking along the beach, just above the tide line, straining my eyes in the dark for the faint discoloration in the sand that would reveal the recent passage of a sea turtle. Usually a companion with a more experienced or sensitive eye than mine would spot it first, and then we would very quietly follow the path up the beach. Sometimes it would prove to be the return path of a turtle that had already nested, or one leg of a half-moon pattern made by a turtle that for some reason aborted its nesting mission. But on occasion, as we ascended to where the beach vegetation began, we would sometimes see but often only hear a female green turtle engaged in her ancient ritual of egg laying. The sound would be the heavy breathing of the turtle. If the turtle was early in the process, still excavating the nest, we remained quiet and at a safe distance from her, because at this time the turtle is alert and if disturbed will abandon her effort. However, if egg laying had already begun, or if the turtle was filling the nest cavity with sand, then we could approach the nest, because now the turtle had entered a trancelike state and was indifferent to our presence. What a thrill it was to kneel behind the turtle and look into the nest and watch the eggs fall one by one into the carefully constructed nest cavity, or to reach down and catch one and then gently lay it on the pile below.

During the entire process, the mother turtle never looks at her nest or at her eggs. All digging, and much of the covering, is performed by her hind legs, which demonstrate remarkable skill and coordination during the excavation process. When she is done, the mother turtle lumbers back down the beach and into the surf without ever looking back. From this time forward, the developing embryos and eventually the emerging hatchlings are completely on their own, with no parental protection or training, relying strictly on their own inborn traits to guide them. The tears running down the face of a nesting turtle have been attributed, by romantically inclined observers, to the sorrow of this separation from her progeny or to the travail of childbirth, although a more objective, scientific, albeit less appealing explanation is that the tears are the products of lachrimal salt glands that serve to eliminate excess salt ingested at sea.

The odds facing these young, innocent embryonic sea turtles incubating in their sandy nest are pitifully slim. Some estimates give only one chance in a thousand of the turtles surviving long enough to engage in reproductive activity of their own. Many will perish within the nest itself if conditions are not compatible with development, or if the nest is dug up by a beach-roaming predator; newly emerged hatchlings will be devoured by ghost crabs, frigate birds, or other beach predators as they scurry down to the sea, and still others will be taken sooner or later by marine predators that prowl the offshore waters. Over the millennia, this seemingly wasteful strategy of egg production has nonetheless served this species well and has always resulted in enough survivors to sustain a healthy population. However, the emergence of our species has sadly tipped the balance. We have not only preyed upon the turtles and their eggs but, more importantly, we have appropriated for our own use most of their historic nesting beaches. Significant conservation efforts have been under way for some time to reverse this destructive process, and we can only hope that it is not too late. (For readers interested in a full discussion of sea turtle conservation and biology, a recent book by James Spotila [2004] is highly recommended.)

Tortuguero

The beach that I had the opportunity to visit on three occasions, Tortuguero Beach, on the Caribbean coast of Costa Rica, is one of the most important of the surviving nesting beaches in the Western Hemisphere. It is preserved as a sanctuary for nesting sea turtles thanks to the efforts of the Caribbean Conservation Corporation, the Costa Rican government, and, above all, the dedication and life's work of Archie Carr, who was a world leader in conservation in general and sea turtle conservation in particular. During his lifetime (1909–1987) Archie was the foremost authority on sea turtle biology. The establishment of the Turtle Research Station and, eventually, the creation of Tortuguero National Park can be traced to his inspiration and dedication. Troëng and Rankin (2005) have documented the great success of this conservation effort for the green turtle at Tortuguero over the past decades.

I had the opportunity to meet Archie during my visits to Tortuguero and, as I am sure it is true for everyone who ever met him, I was charmed by his generosity and sense of humor and amazed by his knowledge of natural history. At Tortuguero he had a deep understanding and compassion for the resident plants, animals, and indigenous people alike. Archie was also a gifted writer, and for lovers of sea turtles I would especially recommend his 1967 book *So Excellent a Fishe: A Natural History of Sea Turtles.*

My visits to Tortuguero were all in the mid-1970s before it became a popular ecotourist destination. Getting there was an adventure, as I gather it still is. The final leg of the trip begins in the Caribbean seaport town of Puerto Limón, about sixty miles down the coast from Tortuguero. A popular place to relax in Limón was the American Bar, situated on the town square. From outdoor seats we could look across the square to the sea, and we were told that Fidel Castro once sat there, perhaps plotting his revolution in Cuba far across the water. Tall trees filled the square, and we were able to see sloths slowly moving about in the canopy. To reach Tortuguero from Puerto Limón, we flew in a small plane and landed on a slender grass strip near the Research Station. The pilots were happy-go-lucky fellows who would often buzz their friends along the way or playfully swoop down to frighten cattle. My companion and co-worker in these expeditions was Henry Prange of Indiana University, who had previously been a colleague of Archie Carr at the University of Florida, and was referred to affectionately by Archie as "El Jefe." Our scientific goal at Tortuguero was to understand the energetics of nesting behavior in green turtles. We timed our visits to coincide with the peak nesting season of this species, which occurs in July and August.

The conditions at Tortuguero were in many ways idyllic: soft, balmy breezes, separation from the realities of the outside world, delicious food, almost always cooked in coconut oil, and stimulating company, usually including Archie. The infrastructure for scientific research, however, was meager at best, and our "laboratory" was an open-sided shelter under swaying palms, with the ever-present danger of being beaned from above by a falling coconut. The lab had no electricity, so on our first trip we relied on instruments that did not have to be plugged in. These included a spirometer, which we used to measure breathing volume, canvas bags, called Douglas bags, which we used to collect expired air, a device called a Scholander 0.5 cc analyzer, which, with some patience and care, could accurately analyze the oxygen and carbon dioxide concentration of a gas sample, a stopwatch, and a generous supply of Imperiale, Costa Rica's finest cerveza.

Research at Tortuguero

This first visit was a relatively brief one during which we studied just two green turtles that we captured on the beach after they had finished covering their nests (Prange and Jackson, 1976). Moving them from the site of capture to our lab was a challenge, as each animal weighed close to 300 pounds (136 kg), so dragging one along the sand was no easy task. To make it easier we tied a long rope to a flipper, let the turtle return to the water, and then, by pulling on the rope, we guided it to the vicinity of the lab, using mostly the turtle's own muscle power (Figure 4.1). Once at the lab, we mounted the turtle on a wooden stand high enough so that its legs could not touch the ground. After an acclimation period we placed an airtight mask on the turtle's head constructed from a large funnel, and then we measured its breathing and metabolic rate.

When a turtle breathed it arched its head upward, as if it were in the water, and exhaled rapidly and audibly. It then immediately inspired and held its breath in the end-inspiratory position until the next breath. For a resting turtle, the pause between breaths was variable but averaged about two minutes. The volume of each breath was large and averaged 4.1 L for a 127 kg animal, some two to four times the volume one would predict for a human scaled to the same body size. This regular pattern of single breaths is similar to the way we breathe, albeit much slower, but it is quite different from the intermittent breathing of freshwater turtles, described in Chapter 3. Both marine and freshwater turtles, however, begin each breath with an expiration and end with an inspiration. This differs from our resting breathing pattern, which begins with an active inspiration and ends with a passive expiration. The respiratory pause of a reptile is thus end-inspiratory, whereas the pause in humans and other mammals is end-expiratory. The initial expiration by a green turtle is clearly a rapid and usually an audible event. How rapid this expiration actually is was quantified by Tenney et al. (1974) on a 100 kg green turtle lying on its back. In this unnatural position, tidal volumes are even larger than in the prone position and can approach the maximal lung capacity of the turtle. Peak flow rates were almost 12 L per second, comparable to what

Figure 4.1   Turtle with its rope "leash" heads for the surf.

an adult human subject achieves during a maximal forced expiration from a fully inflated lung. The larger expired volumes in the supine turtle may have been enhanced by the weight of the viscera pressing down on the lungs. Nonetheless, rapid, large breaths are thought to be important for a marine animal to minimize the time it must spend at the surface. Rapid flow rates also suggest that green turtles have less expiratory flow restriction than occurs in humans during forced expiration. This may be explained by cartilage and smooth muscle in the turtle's terminal airways that help keep the airways open during both expiration and deep diving. These structural features of the airways, a trait shared by marine mammals, permit complete collapse of the lungs during deep diving and thereby avoid excessive nitrogen uptake into the blood at depth and the risk of the bends, or decompression sickness, upon resurfacing (Berkson, 1967; Kooyman, 1973).

Our subsequent visits to Tortuguero were longer and more ambitious. For the first of these, during the following summer, our project required electricity, so we arranged to have a generator shipped to Tortuguero for our use. The principal power-requiring instrument we needed for the study was a clinical-type blood pH and blood gas analyzer that I used in my lab at Brown University. I carefully packed the analyzer into a box and packed other scientific and personal gear in a large steamer trunk that had wheels on one end. When I arrived at the Providence airport for the trip south, I set the trunk upright on its wheels with the analyzer box on top, at about my face level, and I proceeded toward the check-in counter, rolling my cargo ahead of me. In my haste, I lost control of the trunk and it tipped over, sending the box with the precious and delicate analyzer crashing to the floor. I then faced one of those helpless moments in life when you fear that all is lost, but you have no choice but to carry on and hope for the best. My plane was leaving shortly, the success of our expedition depended on a functional analyzer, and I was sure I had destroyed it. It was not a relaxing trip to Miami, and thence to San Jose, over to Puerto Limón, and from there down the coast to Tortuguero. I was a nervous wreck the whole way. However, *mirabile dictu,* when I finally opened the box and tried it out in our lab underneath the coconut palms, the analyzer was intact and worked normally. Thankfully, my packing job was good enough to protect it from my carelessness.

During this second visit, we measured the metabolic rate of green turtles while they were hauling themselves along on the ground, something we had also done in a preliminary way during our first visit. We also succeeded in measuring the metabolic rate of turtles while they were engaged in their natural nesting activity at night (Jackson and Prange, 1979). We of course waited until the turtle was in its trancelike state before we approached it, but once the time was right we could place the mask on it and collect measured expired gas samples just as when it was moving about on the beach. This intrusion did not faze the turtle in the least; it kept on completing its nesting activity as if we were not there. We also were able to sample blood from captive turtles while they were at rest and during exercise to assess blood oxygenation and acid-base status. Once again, all of the turtles we studied were released and crawled back into the sea, perhaps to regale their offshore friends with their bizarre adventures on the beach.

For our final visit two years later, the level of scientific sophistication was raised even higher. Henry successfully secured funding from the National Science Foundation to support a collaborative research mission involving a group of investigators to study the cardiopulmonary physiology of green turtles. For this study we had the services of the R. V. *Alpha Helix*, a well-equipped floating laboratory vessel that sailed out of Scripps Institution of Oceanography in La Jolla, California, and was funded by the National Science Foundation. The ship was the inspiration of renowned animal physiologist Per (Pete) Scholander and had been visiting exotic ports of call for physiological research since her maiden voyage to Cairns, Australia, and the Great Barrier Reef in 1966.

We boarded in Panama City after the ship had returned from a previous expedition to the Galapagos Islands. After traversing the Panama Canal, we cruised up the eastern coast of Central America to Nicaragua, where a number of green turtles were transferred to the ship. Archie, who was on board for this part of the expedition, had arranged for the purchase of turtles from indigenous Miskito Indians, who brought their small boats close to the ship and helped hoist the turtles from their boats onto the *Alpha Helix* (Figure 4.2). With our cargo on board, we returned down the coast to Puerto Limón, where the ship docked for the remainder of the study. Most of our work was done aboard ship, but Henry and I also took an excursion back to Tortuguero to complete our beach studies.

The results that Henry and I obtained from these last two studies confirmed that female green turtles work quite hard throughout the whole process of hauling themselves up the beach, digging the nest, filling the nest, and hauling themselves back into the sea. Their metabolic rate both

Figure 4.2   Local fishermen off the coast of Nicaragua transfer a green turtle from their boat to the R. V. Alpha Helix.

while they "walked" on the beach or when they covered their nest was nine to ten times their predicted standard resting metabolic rate. Standard metabolic rate is the lowest value observed in a quiescent animal that is not digesting any food. As already mentioned in Chapter 3, active green turtles do not apparently exercise and breathe simultaneously. These two behaviors alternate whether the turtle is engaged in nesting activity or hauling about on the beach. The turtle's burst of effort is interrupted by a pause to take several breaths. As discussed earlier, each breath consists of a rapid, audible, deep exhalation followed by an inhalation. The overall frequency of breathing of the active animals was much higher than at rest, but the volume of each breath was unchanged. Our metabolic rate determinations included both the active periods and the pauses, so the metabolic rate during the active phase must have been even higher than the ninefold to tenfold value.

I suggested earlier the possibility that the two functions (breathing and locomotion) employ the same muscles and cannot occur at the same time, although as also discussed in Chapter 3 we now know that freshwater turtles do not have this conflict when they exercise. Another possibility that Henry and I suggested was that each burst of activity exhausted the turtle so much that it had to stop and recover until it could repeat the effort. This is similar to a human athlete engaged in interval training, or "wind sprints," during which periods of intense activity alternate with recovery pauses of deep breathing. Blood samples taken from turtles were consistent with this hypothesis, since the turtles' blood had elevated lactate concentrations, an observation that indicates that the turtles were supporting part of their effort using anaerobic metabolism. We knew previously that many reptiles rely heavily on anaerobic metabolism during bursts of exercise (Bennett and Ruben, 1979).

The entire nesting process of a green turtle lasts some two to three hours from the time the turtle emerges from the water until it returns. Based on our observations and measurements, the overall metabolic rate during this time is as much as nine to ten times the minimum resting rate. This represents a huge expenditure of energy and is comparable to a human athlete running a marathon (42 km) in about the same amount of time, two to three hours. The metabolic rate of a marathoner running at this pace also increases approximately nine to ten times over resting levels (Figure 4.3). Anyone who has ever run a marathon can attest to the deep fatigue experienced at the end of the race. Recall though that the turtle may repeat this endeavor several times each nesting season, so it

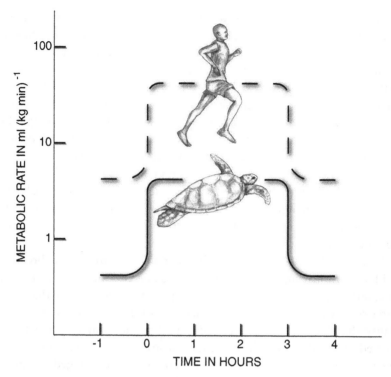

Figure 4.3 The increase in metabolic rate during nesting by a green turtle is comparable in duration and relative magnitude to the energy expenditure of a marathon runner. The absolute value of the turtle's metabolism is, however, about ten times lower.

may exert physical effort equivalent to perhaps four marathons in the course of six weeks or so.

Sea Turtle Migration

The effort expended during nesting is not the only energetic challenge these sea turtles experience. The waters offshore from a nesting beach such as Tortuguero are not the waters where the turtles feed. The feeding areas (the turtle's diet consists here of sea grass and algae) are generally far removed from the nesting beach (Broderick et al., 2001). Based on the Tortuguero tagging records, turtles nesting there may travel hundreds of kilometers in the Caribbean to feeding areas along the coasts of Venezuela, Cuba, and the Yucatan. The nesting turtle probably goes without food from the time it leaves its feeding ground and swims the many miles between there and the nesting beach until it returns to its feeding area some months later.

An extreme example of long-distance migratory behavior is a population of green turtles that feed off the coast of Brazil but build their nests and lay their eggs on Ascension Island, some 2,000 km away across the Atlantic Ocean (Carr, 1975). The elapsed time between leaving the feeding area and returning there must be greater than four months, during which time the turtles may never feed. Turtles on their feeding ground must accumulate a great deal of stored fuel in the form of fat before undertaking a starvation voyage of this magnitude. This probably helps explain why they do not make the trip every year. My colleague Henry Prange measured the metabolic rate of young green turtles (< 1 kg) swimming in a flume and, based on these data and some reasonable extrapolations, he estimated that an adult turtle making the round trip between Brazil and Ascension Island would require an amount of fat equivalent to 21% of its entire body weight (Prange, 1976). This may seem like a lot of fat, but birds can have fat stores equal to 50% of their body weight prior to migration. Henry's estimate is probably low, however, because his calculation did not include the additional cost for the turtles offshore at the nesting beach and during the vigorous excursions onshore to lay eggs. He also did not take into account the possibility of ocean currents either helping or hindering their progress. Despite these uncertainties, the process of nesting by sea turtles is without question a remarkable physical feat.

Why do the Ascension Island green turtles, which number in the thousands nesting each year, travel so far and to such a remote island in order to nest? The answer is uncertain, but Archie Carr and his colleague Patrick Coleman proposed a provocative explanation some years ago (Carr and Coleman, 1974). Their idea was that this pattern began tens of millions of years ago during the late Cretaceous period, when Africa and South America began to move apart from each other due to sea floor spreading. Protected areas on the South American coast provided sites for feeding, whereas emergent volcanic islands offshore provided appropriate sites for nesting. Exposed shores such as those that occur on islands provided the high-energy surf conditions optimal for nesting, plus historically they lacked the numerous terrestrial predators that threaten continental nesters. Volcanic islands tend to emerge along mid-oceanic ridges, and so the ancient predecessors of our contemporary green turtles may have begun nesting on islands off of the South American coast and continued this pattern over the thousands of millennia as the continents drifted farther and farther apart and as old islands sank and new islands arose. According to this theory, Ascension Island, at just several million years old, is the latest in a succession of such islands that formed near the ridge. One seemingly unlikely assumption for this idea is that there was never a significant interval when no island with suitable nesting sites emerged above the ocean's surface. Given the habits of green turtles (as we know them today) to return to their native beach, an absence of an island for even a couple generations of turtles could wipe out the population.

More recently Bowen et al. (1989) challenged the Carr–Coleman hypothesis. They reasoned that if the Ascension Island population has been isolated from other populations of green turtles for as long as 40 million years, then considerable genetic drift would be apparent in this population. They chose to examine mitochondrial DNA (mtDNA), which is always transmitted to the offspring from the mother. They compared mtDNA from eggs or hatchlings collected from three sites in the Atlantic (including Ascension Island) and one site in the Pacific. Considerable drift was found between the Atlantic and Pacific populations, indicating a long period of isolation from each other, but much smaller differences were observed in the Atlantic population. They concluded that the colonization of Ascension Island must have been a rather recent event and not an isolated population with an unbroken history stretching back tens of millions of years. Contemporary tagging records indicate that female turtles occasionally nest on different beaches separated by long distances, so site fidelity is not invariant. If Bowen and colleagues' conclusion is correct, and if Carr and Coleman's intriguing idea is incorrect, then it still remains a mystery as to why a major nesting colony would be established so far from the feeding grounds.

A remarkable aspect of these periodic migrations of green turtles and other sea turtles is their uncanny ability to find their way to their destination. Imagine swimming 2,000 km across a vast ocean and homing in on Ascension Island, a roughly circular outcropping scarcely 10 km in diameter. The current hypothesis is that turtles use their established ability to detect the earth's magnetic field to construct a magnetic map of their destination that leads them to the vicinity of their goal, and then they use nonmagnetic senses, such as vision and olfaction, to find the exact location (Lohmann et al., 2008).

An important conservation-related aspect of the green turtle's migratory habits is that the turtles most often return to the same area to nest. Furthermore, not just the females swim all the way from the feeding grounds; the males also make the trip. Mating occurs in the waters off the nesting beach. Because the members of a single population, such as those that nest at Ascension Island or Tortuguero, almost invariably nest at the same location, various populations of green turtles must be sexually isolated from each other. Consequently, turtles from one population are unlikely to repopulate another nesting area where that population has disappeared, although given the Ascension Island example, it is not impossible. Conservation thus should consider all populations distinct and potentially threatened, regardless of the worldwide numbers. Repeated fidelity of nesting turtles to return consistently to the same site may be traced back to these turtles having hatched at that site, although this is difficult to prove. Assuming that adult turtles do return to their natal beach to mate and nest, this behavior offers the possibility of collecting turtle eggs at a well-populated beach and incubating the eggs on a different beach where the population has dwindled. The idea is that the hatchlings will imprint somehow on the beach where they emerge and when mature will return to the very same beach for mating and nesting. The problem is that twenty-five years or so must elapse before a green turtle becomes sexually mature and is ready to reproduce, thus the results of such a study are hard to confirm.

The reproductive biology of sea turtles, as well as other turtles, has long been a source of fascination, perhaps expressed best by poet Ogden Nash:

> The turtle lives 'twixt plated decks
> Which practically conceal its sex.
> I think it clever of the turtle
> In such a fix to be so fertile.

The anatomy of the male turtle helps overcome the challenge of the "plated decks." His strong foreleg claws enable him to tightly grip the front of the female's shell, and his long tail (a common diagnostic trait of a male) enables him to extend it around the back end of the female's shell. His protruded penis can then inseminate the female. It may sound easy, but for sea turtles and freshwater turtles, this activity occurs in water, and the struggling pair must have access to the surface for breathing. Additionally, the female may resist the male's advances, and other males may hover about awaiting their opportunity. Terrestrial turtles may have a somewhat easier time, but conditions must be right for them as well. I once visited the reptile house at the Philadelphia Zoo at a time when the giant Galapagos turtles were being temporarily housed on a cement floor while new reptile quarters were being built. I watched, in embarrassed amusement, as the frustrated male attempted unsuccessfully to mount his beloved because he lacked the soft dirt surface into which he could dig his hind legs.

In the waters off of Tortuguero Beach, the gathering of the green turtles is a promiscuous affair during which a female may copulate with several males, and each male may inseminate a number of females. As a result, a clutch of eggs from a single mother may contain genes from several different fathers. This practice is common among both sea turtles and freshwater turtles (Pearse and Avise, 2001), although available evidence indicates that the female leatherback sea turtle may restrict her affections to a single favored male. In contrast, a female may not need to mate at all during a given year but instead may utilize sperm stored in her oviduct from a previous mating engagement, perhaps from more than a year ago. This practice of storing sperm for later use is common among turtles generally. A suggested advantage of this strategy is that it enables the female to fully exploit the sperm of a particularly fit male or minimize the cost of a potentially demanding mating activity.

Temperature-Dependent Sex Determination

The sex of a human baby is genetically determined. If the baby inherits an X chromosome from each parent, then it is female; if instead the baby gets an X from its mother and a Y from its father, then it is male. However, this is not a universal rule among vertebrates. In most turtles and many other reptiles a fertilized egg, after it is laid, has the potential to develop into either a male or a female hatchling. The newborn's sex depends on the incubation temperature, a phenomenon called temperature-dependent sex determination (TSD). In a study conducted at Tortuguero, Morreale et al. (1982) collected clutches of eggs from nesting green turtles and then reburied them at sites selected to produce different thermal environments. Five clutches of about one hundred eggs each were used for each of four temperature regimes. The researchers, after the egg hatching, collected a sample of the young turtles for sexing and released the rest. Their results were quite clear. Eggs that incubated at warm temperatures above 29.5°C developed predominately into female hatchlings, whereas eggs incubated at cool temperatures below 28°C developed mainly into male hatchlings. Similar results have been found in studies of other green turtle populations; the pivotal temperature at which the sex ratio is approximately 1:1 appears to be approximately 29°C. The sex of any individual hatchling can thus depend on a number of factors that influence the incubation temperature. These include the location on the beach where the nest is built, where within the nest the egg is situated, when during the nesting season the egg is laid, and what the weather happens to be during the middle third of development, the critical period when sex is determined. It is worth noting, however, that the reliability of sex determination in sea turtle hatchlings using non-destructive methods has been questioned and remains a challenge for scientists studying this issue (Witt et al., 2010).

A long-standing question has been what adaptive advantage is there for an animal utilizing TSD? That is, is the fitness of a male improved because it experienced a lower temperature during its incubation, or does a female have more success in life because of a higher developmental temperature? This is not an easy question to answer, for a couple of reasons. First, to compare fitness or reproductive success, it is necessary to compare females and males at both high and low temperatures. However, in turtles, generally only males are produced at low temperature and females at high temperature. Second, many reptiles utilizing TSD have very long life spans and, like green turtles, they may not reach reproductive age until they are twenty to thirty years old. (Try getting research money from the National Science Foundation for a project that lasts thirty years or more.)

Warner and Shine (2008), who conducted their recent study in Australia, managed to find a way around both of these experimental obstacles. For their experimental animal they selected a lizard, the jacky dragon *(Amphibolurus muricatus)*, which produces primarily female offspring at both low (23°–26°C) and high (30°–33°C) incubation temperatures and a mixture of male and female offspring at intermediate temperatures. The short life span of this lizard permitted these investigators to complete their study of reproductive success within the funding agencies' approved time scale of three to four years. Because females of this species are produced at all temperatures, no manipulation was required to test the temperature-dependent fitness of this sex. The challenge was how to produce males at high and low temperatures.

Here is where understanding the basic biochemistry of TSD provided the answer (Crews et al., 1995). In TSD species generally, testosterone serves as a precursor molecule that can either be converted to estradiol by an enzyme called aromatase or into dihydrotestosterone by a reductase enzyme. Female-favoring temperatures lead to the production of aromatase and estradiol, whereas male-favoring temperatures lead to the production of reductase and dihydrotestosterone, producing, respectively, female and male offspring. Warner and Shine (2008) incubated clutches of jacky dragon eggs at high, low, and intermediate temperatures and treated half of the eggs in each clutch with an inhibitor of aromatase. Eggs treated with the inhibitor developed into male offspring, regardless of the incubation temperature, thereby surmounting the second obstacle—offspring of each sex were now produced at all three incubation temperatures. These baby lizards then grew to maturity, and the scientists monitored their success at producing offspring. The important finding was that the male lizards that developed at the intermediate temperature of 27°C, the temperature at which males normally develop, were significantly more successful at producing offspring than the males that developed at either low or high temperature. Although the result was not as dramatic, females that developed at extreme temperatures produced more offspring in their lifetimes than those that developed at the intermediate temperature. The conclusion was that reproductive fitness is improved when development occurs at the appropriate temperature for each sex.

One potentially serious implication of TSD that Morreale and his colleagues (1982) emphasized in their paper relates to the practice of artificially incubating eggs for the purpose of conservation. Such a conservation strategy could be detrimental if those managing the incubation conditions were not informed about the importance of incubation temperature. Extremes in either direction could produce either all males or all females, not a good outcome for the population. A more critical problem, however, is global warming. Rising temperatures may pose a very serious threat to green turtles and other species that exhibit TSD (Wit et al., 2010). As temperatures rise at the nesting beaches, a shift toward increasing numbers of female hatchlings will occur, threatening the extinction of the species. Studies that have measured the relative abundance of male and female hatchling green turtles at various nesting beaches around the world have generally agreed that even now a preponderant number of female hatchlings are produced (e.g., Wibbels et al., 1993; Broderick et al., 2000). This suggests that it would not take a major increase in global temperature to tip the balance completely over to all female progeny. This sad outcome, if it should occur, may not be without precedent in the earth's history. One theory for the extinction of the dinosaurs is that falling temperatures at the end of the Cretaceous period led to the production of single-sex offspring and the eventual end of those iconic reptiles.

An intriguing question is how much control, if any, nesting female green turtles have over the sex of their offspring. Certainly observing their behavior on the beach makes it clear that they are quite particular about where they build their nests. Might they have some signal that makes them, for example, more likely to choose a cooler site so that male hatchlings would result if the number of males in the population were declining? A beach such as Tortuguero does offer a variety of sites that will have quite different temperature profiles during the incubation period. If so, then the turtles might be able to stave off the effects of a warming climate by their judicious selection of nesting sites and thereby keep the males coming. Of course, our species, arguably the prime culprit in global warming, could intervene by collecting eggs and incubating them at temperatures appropriate to sustain a healthy sex ratio. This could be a formidable task, however, not only because of the distribution and size of the green turtle population but also because other species of sea turtles, plus freshwater turtles and other reptiles, may face the same reproductive dilemma. The best approach would be to do all we can to arrest global warming through a massive campaign to reduce greenhouse gas emissions.

5

OVERWINTERING WITHOUT BREATHING

The time is late fall in central Wisconsin. The sun has nearly gone down, the sky is clear, and the temperature is falling. A painted turtle that lives in a quiet pond swims to the surface, takes several breaths, raising the level of oxygen in its lungs, and then sinks back into the water. The water is already so cold that the turtle's metabolism has slowed to a small fraction of what it is in the summer, and the turtle may not have to come up again to breathe for many hours. When it does try again to breathe, it finds its access to the surface blocked by a sheet of ice. It searches for an opening but finds none. It returns to the bottom and waits—and waits. It may explore the surface again for a breathing hole, but the surface ice gradually thickens. Winter settles in and many weeks may pass before the ice melts and the turtle can take another breath of air.

It is not easy for us to imagine what this experience is like: in almost total darkness except for faint light that may penetrate the ice and snow cover on the pond during the brief daylight hours, in extreme cold that slows down to a minimum all of its living processes, the turtle sits quietly, scarcely moving, with only a minimal level of consciousness. Yet for a painted turtle living in the northern part of its range, as much as half of its total lifetime may be spent in this state.

Although the description I have just given is a romanticized and not scientifically documented account of hibernation in a painted turtle, some variation of this dramatic scenario plays out each winter for freshwater turtles living in the temperate zone in North America. A protracted period of breath holding is an experience a turtle must somehow endure if it is to see another spring. How does it do this? Like us, turtles rely ultimately on oxygen for life. Without oxygen they cannot develop, grow, process food, and reproduce. Unlike humans and most other air-breathing vertebrates, however, they can survive for a remarkably long time without breathing.

Breath Holding in Humans

To appreciate what the turtle can do, it may be helpful to relate its wintertime experience to our own physiology. With your eye on a clock with a second hand, take a deep breath and hold it and see how long it takes before you start to feel very uncomfortable. If you are like most people, you will grow so desperate for air that you will give up and start breathing again in about a minute. Something very powerful happened during that minute to produce an irresistible urge to breathe.

Breathing is something we do without conscious effort. Day in and day out, we breathe rhythmically, about twelve to fifteen times every minute. If we are active, then we breathe a bit faster and more deeply. The additional air supplies the higher demand our tissues have for oxygen and keeps the gas composition of our lungs and blood, specifically the concentrations of oxygen and carbon dioxide, constant. This is an example of homeostasis, the maintenance of a relatively constant composition of the fluids in our body. For breathing, it is a complex balance between the rate at which oxygen is used and the rate at which it is replaced and a balance between the rate at which carbon dioxide is produced and the rate at which it is lost. If these balances are disturbed, as they are when you hold your breath, then control systems spring into action to correct the imbalance and restore the homeostatic state. Included in the control response is a strong, conscious urge to breathe.

It is possible to hold your breath longer if you breathe rapidly and deeply before you start holding your breath. This maneuver, called hyperventilation, drives your lung oxygen up and carbon dioxide down so that it takes longer for your metabolic consumption of oxygen and production of carbon dioxide to bring the levels of these gases back to the point where they stimulate your breathing again. How long you can hold your breath is ultimately limited, though, by how much oxygen you have in your body and by how fast your body's cells are using up the oxygen. We cannot survive long without oxygen, because once the oxygen is gone, so are we. It turns out that this simple relationship, how much oxygen we have available and how fast we use up the oxygen, places a rather strict upper limit on our maximal breath-holding time. Most of the oxygen in our body at the beginning of a breath hold is in our lungs and in our blood bound to the hemoglobin within our red blood cells. The magnitude of this oxygen reserve for an average-size adult totals only about 1.5 L of oxygen. This same average person at rest consumes approximately 0.3 L of oxygen per minute to support whole-body oxidative metabolism. A simple calculation reveals that at this rate of consumption, all of the body's oxygen will be gone in five minutes. Well before you reach that limit, though, your control systems compel you to breathe. By sheer willpower and some practice, many individuals can resist the urge to breathe and extend their breath hold time two to three minutes, but the upper limit defined by our oxygen reserves still exists. I observed this in a student laboratory that I taught at Brown University. Student subjects held their breath for varying periods of time while at rest. At the end of each breath hold, a sample of their lung gas was collected and analyzed. As shown in Figure 5.1, their longest breath holds were only about one to two minutes and, assuming they could have kept going until all of their oxygen was gone, they could have lasted, as predicted earlier, only four to five minutes.

Remarkably, the five-minute limit can be exceeded. Free diving is an international sport with competitions and records in various categories. It is performed strictly with breath holding and not with the use of diving apparatus such as scuba gear. One category of competitive free diving is called static apnea, apnea being the scientific term for not breathing, that is, breath holding. The static apnea competitor prepares by hyperventilating and filling his or her lungs as much as possible with air (no pure oxygen allowed). He or she then submerges, with nose and mouth underwater, and holds his or her breath as long as possible. The recent world record for static apnea is an amazing 11:35, by Stéphan Mifsud from France. His performance is compared to the students in my laboratory class in Figure 5.1.

How is this incredible breath hold possible given the limitation defined earlier? There are only two ways Mifsud could accomplish this given the safe assumption that he could not survive for several minutes without oxygen in his body. First, he can have more oxygen on board at the beginning of the breath hold than most of us do. He could accomplish this by having a larger-than-normal lung volume and/or by having a larger-than-normal blood volume with a higher-than-normal concentration of red blood cells. This would allow his body to contain more than the 1.5 L of oxygen considered average. Second, he may be able to reduce his metabolism below the normal resting level by relaxation techniques or meditation and thereby slow the rate at which he uses up his oxygen stores. Many people can hold their breath two minutes or more, so if you could double your oxygen stores and halve your meta-

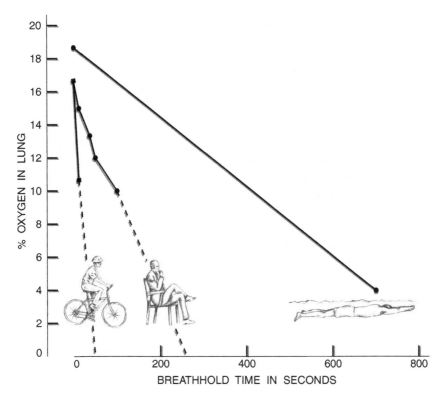

Figure 5.1 Lung oxygen of student volunteers decreases during breath holds at rest and during mild exercise on a stationary bicycle (solid lines), but the changes occur much faster during exercise. The dashed lines are hypothetical extensions to the zero oxygen line to estimate how long the lung oxygen would last under each condition. Note that the world record breath-hold duration to the right far exceeds the predicted extreme limit for the students at rest. Oxygen values for the record breath hold were not actually measured, and the values shown are estimates based on a published study of competitive divers (Lindholm and Lundgren, 2006).

bolic rate, then the world record would be within reach. This may sound easy, but achieving what Mifsud did is extraordinary and had to require training, practice, probably some natural ability, and very strong willpower. The rest of us can only marvel at his performance and realize that even with his level of training and effort we would still fall far short of the standard he has set.

Breath Holding in Elephant Seals

However, eleven-plus minutes still pales in comparison to the months of breath holding by the humble turtle. It also falls well short of various aquatic mammals and birds whose normal lifestyles involve regular excursions beneath the water's surface to feed and carry out other activities. Probably the most exceptional animal on which we have detailed information in this regard is the elephant seal, a wide-ranging resident of the Pacific Ocean. Elephant seals are earless or "true" seals, members of the family *Phocidae*. Unlike eared seals, or *Otariadae,* such as sea lions, they are rather clumsy on land but are adapted for long-distance swimming with powerful hind limbs. Elephant seals spend long periods at sea but come ashore at predictable times for reproduction and molting. This life-history pattern makes the elephant seal well suited for studies on the behavior and physiology of a free-ranging marine mammal. Devices, securely attached to the skin of seals while on land, can record data while the animals are at sea and can then be recovered when they return to the same location weeks later. A well-studied population of northern elephant seals, *Mirounga angustirostris,* has a rookery at Año Nuevo Point in California. Experimental studies on these seals have revealed extraordinary diving behavior, including the frequency, duration, and depth of dives. A groundbreaking study of elephant seals by LeBoeuf and coworkers (1988) using time-depth recorders documented that the seals routinely dive around the clock for weeks at a time with only brief periods of breathing at the surface. Individual dives average about twenty minutes but can last an hour or more, and the seals regularly reach depths of 400–800 m (Figure 5.2).

Seals and other diving mammals and birds, like human breath holders, are ultimately limited by the rate at which they deplete their oxygen stores. All warm-blooded vertebrates require the continuous availability of oxygen to their cells for survival, particularly because of the sensitivity of the heart and central nervous system to the lack of oxygen. Elephant seals are no exception to this rule. How, then, can we explain the superior performance of elephant seals compared to humans, even to champion breath holder Mifsud? The explanation must go back to the basic relationship between the size of the body's oxygen reserve and the rate at which it is used. A female elephant seal that weighs 333 kg, about five times the size of a typical adult human, has an estimated total oxygen store of 24 L. Recall that the human oxygen store is about 1.5 L. By scal-

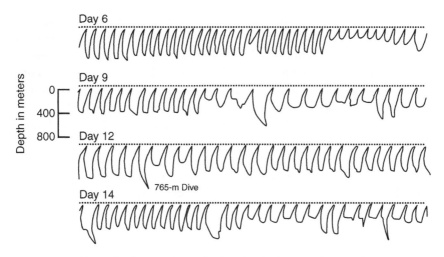

Figure 5.2 Time-depth recording of a female elephant seal at sea diving continuously. The interval between dots above the records is twelve seconds. This diving pattern goes on, around the clock, for weeks at a time (from LeBoeuf et al., 1988).

ing up the human value to the larger elephant seal, we would predict that a seal would possess a reserve of only about 7 L, less than a third of the actual value. This means that, per kg, a seal has over three times as much oxygen as a human. Most of the seal's oxygen is located in its blood, and both the volume of blood and concentration of oxygen-carrying red blood cells are exceptionally large in this animal. Recent work by Meir and colleagues (Meir et al., 2009) has also revealed that elephant seals, diving freely in the open ocean, can utilize nearly all of the oxygen in their blood before returning to the surface to breathe. The authors speculate that the seals must have specialized protective measures to sustain brain function at such low oxygen levels.

In addition, the resting, or standard, metabolic rate per kilogram of body mass of a seal is lower than in a human, in part because of the well-established relationship between metabolic rate and body mass. A large animal typically has a lower weight-specific resting metabolic rate than a small animal. A familiar and dramatic example is a 4,000,000 g elephant, which, not surprisingly, has a total metabolic rate that is much larger than that of a 25 g mouse; but, less predictably, each gram of mouse has a metabolic rate that is nearly twenty-five times as great as each gram of elephant.

This size dependence of metabolic rate holds true for mammalian species generally. The difference in size between the seal and the human is less than that between the elephant and the mouse, but on the same basis, our metabolic rate per gram, that is, the rate at which each gram of our body consumes oxygen, is 50% larger than that of an elephant seal. As a consequence of these two traits of the seal, a larger oxygen store and a slower depletion of that store, the seal should be able to hold its breath longer than we can.

However, there is another factor that adds to the seal's advantage—the diving reflex. As a dive begins, the heart rate slows and the heart pumps less blood; the blood that it does pump, though, distributes preferentially to the heart and brain, the organs that are most dependent on a continuous supply of oxygen. Less vulnerable organs, such as the kidneys, skin, and gastrointestinal tract, receive less blood, and their oxygen consumption falls. Blood flow to the muscles also slows down, but an elephant seal's muscles have a high concentration of myoglobin, an oxygen-binding molecule similar to the hemoglobin of blood, which provides the muscles with their own private oxygen reserve. We can now appreciate how a run-of-the-mill elephant seal can routinely hold its breath for twenty minutes or more, whereas even our world champion breath holder's time is less than twelve minutes (and few of us can exceed three minutes).

Recall as well that our world champion is holding his breath while at rest, and indeed he is probably in a hypometabolic Zen state. In contrast, the seal is actively diving to depth and back and making its living underwater. When leaving the rookery after weaning her pup, a female elephant seal is emaciated because she has not eaten for over a month and during this time has used her body stores to support not only her own metabolic needs but also to produce the protein- and fat-rich milk for a rapidly growing seal pup. When she finally returns to the sea, she must be famished. The study by LeBoeuf and colleagues (1988) found that during the ensuing ten weeks at sea after leaving the rookery, the seal gained weight, on average, by more than 1 kg per day. By the time the seal returned to the rookery to molt, its weight had increased by nearly 25%. To put this in perspective, if you began a ten-week period of binge eating with a body weight of 160 pounds, then you would weigh in at 200 pounds at the end, an impressive weight gain. Although during its very long dives the seal may be resting or sleeping, the active foraging for food must have occupied many of its dives, making it all the more remarkable that the seal can repeatedly dive while holding its breath for so long.

What happens when humans exercise while holding their breath? In the same laboratory exercise I described earlier, we also measured the lung oxygen concentration after the students had held their breath for fifteen seconds while riding a stationary bicycle at a very comfortable pace. Based on this measurement, we calculated that lung oxygen would be exhausted in less than a minute (Figure 5.1). A rather mild level of exercise had drastically shortened the maximal breath-hold duration. In contrast, the diving records of the elephant seal strongly suggest that this animal can be active while holding its breath and yet remain submerged and aerobic for twenty or more minutes. It remains uncertain how they are able to do this.

Breath Holding in Overwintering Turtles

An elephant seal is, without question, a champion mammalian diver, yet even this highly adapted animal cannot approach the performance of an overwintering freshwater turtle. In fairness to the seal, some key differences tilt the balance in the turtle's favor. A turtle is a cold-blooded animal, so its body temperature in the cold winter pond is essentially the same as the water, whereas the warm-blooded elephant seal maintains a temperature in the range of 35°–37°C, even in cold ocean water. The cold temperature depresses chemical reactions in the body, so the turtle's cold body temperature greatly reduces its metabolic rate, which is already low because the turtle is a reptile. Unlike the seal, an overwintering turtle is minimally active, although some moving about does occur. The rate at which a turtle under these conditions consumes oxygen is decidedly slow, but how long will the oxygen stored in its body last once it stops breathing and sinks to the bottom of a frozen pond? Based on the estimated oxygen stores of a painted turtle, of which about two-thirds are in the lungs and one-third in the blood, we can calculate that the oxygen would last about twenty-one hours. The is an impressive breath hold duration compared to a human, or even to an elephant seal, but it is well short of the months that a turtle can actually remain in this state.

However, another significant difference exists between the diving mammal, whether human or seal, and the freshwater turtle. The only available oxygen a diving mammal has is what is in its body when it goes underwater, because its large size and impermeable skin preclude any meaningful oxygen uptake from the water. A much smaller freshwater turtle, on the other hand, can take up oxygen directly from the water and can satisfy all or most of its cellular requirements in the winter pond if the water is well oxygenated. This is despite the obvious fact that a turtle, compared to other aquatic cold-blooded vertebrates, such as fish, with their gills, and amphibians, with their permeable skin, appears quite ill suited for aquatic gas exchange. Finally, as will be discussed fully in Chapter 6, a turtle has another advantage not shared by seals or even by most other cold-blooded animals. It can survive with no oxygen whatsoever, in its body or in the water surrounding it.

Aquatic Oxygen Uptake—General Principles

Obtaining oxygen from water is no easy chore, even for an animal like a fish that does it full time. Oxygen is poorly soluble in water, which means that a volume of water equilibrated with air contains far fewer oxygen molecules than the same volume of air. Even in cold, air-equilibrated water, in which oxygen solubility is relatively high, the concentration of oxygen molecules is only about 5% of the concentration in sea-level air. Therefore, when a fish moves a liter of water through its gills, only 10 mL of oxygen will be available for uptake, whereas a liter of air provides over 200 mL of oxygen to the air breather. Even at the summit of Mount Everest, where a nonacclimatized person could not survive because of the low oxygen, a liter of air still has more than five times as much oxygen as does a liter of sea-level pond water. Based on this comparison, it may seem amazing that an animal can survive at all when breathing only water, even an animal such as a fish, with its highly effective gill exchange surface.

However, the situation on Mount Everest differs from the situation in sea level water in one very important respect relevant to gas exchange. The pressure exerted by the oxygen molecules is higher in the water. At high altitude, as at sea level, oxygen accounts for about 21% of the total gas molecules, and oxygen therefore exerts about 21% of the total gas pressure; however, the total pressure, the atmospheric pressure, is much lower at high altitude. On the summit of Mount Everest the partial pressure of oxygen in the air that a climber breathes into his or her lungs is only about 43 mmHg (West et al., 1983). In pond water equilibrated with sea-level air, in contrast, oxygen partial pressure is the same as it is in air, approximately 150 mmHg, because oxygen diffuses and equilibrates on the basis of partial pressure. Considering only oxygen partial pressure, therefore, the sea-level air breather and the sea-level water breather are on equal footing. By moving enough water through its gills, a fish can raise the oxygen partial pressure in its blood to as high a level as can an air breather. However, to do so, the fish must breathe some twenty times the volume of an air breather to obtain the same amount of oxygen. On Mount Everest, no matter how much air you breathe, you can never raise the oxygen partial pressure in your blood higher than 43 mmHg.

A turtle does not have gills, though, and it lacks the permeable skin that enables amphibians, such as frogs and salamanders, to satisfy much of their gas exchange requirements even at summer temperatures by direct exchange with water. With the exception of certain turtle species, such as the softshell and the Australian Fitzroy River turtle, most turtles can supply only a small and an often negligible fraction of their oxygen directly from water at summer temperatures. Normal gas exchange requires periodic trips to the surface to breathe and to extract oxygen from the air.

How, then, can they rely more fully on aquatic gas exchange at a low temperature? The reason is that a very cold temperature profoundly depresses the turtle's metabolic rate but has much less of an effect on the rate of oxygen diffusion. As a result, rates of gas diffusion that are inconsequential in the summer can be adequate in the winter. Oxygen consumption of a painted turtle at 3°C with access to air is less than 6% of the rate at 20°C (Herbert and Jackson, 1985b) and, as will be discussed later, breath holding may cause an even further decrease in metabolism. In contrast, the rate of gas diffusion, the principal way that oxygen moves from the water into a submerged turtle, decreases much less at 3°C and is still about 85% of the 20°C rate. Consequently, uptake of oxygen from the water can supply an increasingly large fraction of the turtle's oxygen requirements as temperature falls.

Oxygen Uptake by Submerged Turtles

In some species, enough oxygen can be taken up by the skin to provide all of the turtle's oxygen requirements at temperatures near the freezing point of water. One such species is the map turtle, *Graptemys geographica,* a freshwater turtle that is found in the east-central United States, extending up into Wisconsin, Michigan, and northern New England. A population of map turtles resides in the Lamoille River that flows into Lake Champlain in northern Vermont, and members of this population overwinter communally in this river. Preliminary laboratory studies on individuals from this population revealed that they could remain submerged at 3°C in aerated water for many weeks with no significant increase in blood lactate concentration (Ultsch and Jackson, 1995). This observation is significant because it is strong evidence that the map turtles were able to obtain all of the oxygen they needed directly from the water. Had this uptake been insufficient, the turtles would have had to supplement their energy production with anaerobic metabolism, and this would have produced lactic acid, the metabolic end product of anaerobic metabolism. At the pH of the body fluids, some 99.9% of the lactic acid dissociates into the anion lactate and a proton (H^+), so the concentration in blood is normally expressed as lactate concentration, plus technically it is the anion lactate that is produced by the anaerobic pathway, not lactic acid (Hochachka and Mommsen, 1983). The low concentration of lactate in the blood of the map turtles revealed that these turtles were fully aerobic while submerged at a low temperature.

Based on this observation and the existence of an "accessible" population of hibernating map turtles, a project directed by Gordon Ultsch at the University of Alabama was launched to study these turtles on site, in the Lamoille River in northern Vermont. An intrepid postdoctoral fellow, Carlos Crocker, who was an experienced scuba diver and who was working in my lab at the time, agreed to carry out this study. With colleague Terry Graham of Worchester State College providing support at the surface, Carlos made a series of dives into the frozen river over the course of a full winter, collecting several turtles on each dive and obtaining a blood sample from each turtle while it was still underwater. These blood samples were then transported to a nearby lab for analysis of blood variables that are diagnostic of the metabolic state of the animals, including pH, lactate concentration, and oxygen partial pressure. Remarkably, the results Carlos obtained revealed that the turtle's blood pH barely changed from month to month, and lactate rose only slightly, confirming the effectiveness of this turtle's aquatic respiration (Crocker, Graham, Ultsch, and Jackson, 2000). The oxygen partial pressure in the river remained high, near the atmospheric oxygen pressure, throughout the winter, and the turtles were always sitting on the bottom, fully exposed to the water.

The logistical challenges of this study should not be overlooked. Northern Vermont is extremely cold in the winter, and the ice cover on the river was some 30 cm (12 inches) thick. Carlos used a chain saw to create a hole in the ice large enough to allow him to enter the water. A rope connected him to his colleague at the surface so the colleague could assist him in his return to dry land and be available to help with any emergency. Diving into water under these conditions is challenging enough, but Carlos also located the hibernating turtles, collected three of them on each occasion, and obtained a blood sample from each without letting it breathe. In the interest of full disclosure, it should be noted that Gordon Ultsch and I, the senior authors in this study, remained in our warm home quarters during Carlos's monthly visits to northern Vermont.

Most of the map turtles studied were females, who tipped the scales at 1–2 kg. Males of this species are much smaller, weighing in at a paltry 0.1–0.2 kg. An adult female map turtle is an unlikely animal to be so successful at aquatic gas exchange. It is a relatively large turtle with a hard, mineralized shell and thick, reptilian skin, surfaces poorly suited for respiratory exchange. It is possible that the map turtle can supplement its limited skin function by pumping water in and out of its mouth or in and out of its cloaca to exploit other, more permeable surfaces for acquiring oxygen. It may also be able to reduce its metabolic rate even more than other species so that the amount of oxygen needed is minimized. We simply do not know. What we do know is that whether sitting beneath the ice of a Vermont river or in an experimental tank filled with cold, aerated water in the laboratory, this turtle can extract enough oxygen from the water to meet the demands of its tissues.

Other turtle species with northern distribution, such as the musk and softshell (Ultsch and Cochran, 1994; Reese et al., 2003), can also remain aerobic during winter hibernation, but these types seem better suited for the job than the map turtles. The musk turtle, *Sternotherus odoratus,* is a small animal that weighs approximately 100 g, so it has a favorable relative surface area. In addition, its exposed skin surface is relatively large due to the small size of the ventral shell (the plastron), and folds on the surface of its skin add to the area for gas exchange. The softshell, *Apalone spinifera,* is a larger turtle but with a shell much less mineralized than that of other turtles and covered with a leathery skin that provides a pathway for gas exchange. The softshell also has specialized papillae in its oral cavity that provide an expanded surface for gas exchange when the turtle moves water in and out of its mouth (Wang et al., 1989). This turtle often sits on the bottom of a pond concealed under a layer of sand or mud with its snout extended up so that it can "breathe" the water. Both the musk and the softshell can satisfy their oxygen needs even in water as warm as 10°C (Ultsch et al., 1984), and at winter temperatures, 3°C or less, they can remain submerged in aerated water for many months with no evidence of anaerobic metabolism.

In the study of the softshell in aerated water at 3°C (Reese et al., 2003), blood samples collected periodically over the 150 days of submergence were analyzed not only for lactate concentrations but also for pH, blood gases, and various ions. The measurements of blood PCO_2 were particularly informative and provide strong circumstantial evidence for metabolic depression during submergence. Prior to submergence, when the turtles still had access to air, blood PCO_2 was about 10 mmHg. By the tenth day PCO_2 had fallen to 6 mmHg, and by the twenty-fifth day it was only 4 mmHg, where it stayed for the remainder of the submergence period. In an animal exchanging CO_2 strictly by diffusion, blood PCO_2 should vary directly with the turtle's metabolic CO_2 production and inversely with the diffusing capacity for CO_2. Therefore, the decrease we observed could have been due either to an increase in diffusing capacity or to a decrease in metabolic CO_2 production. Assuming that the latter change is more likely, the observed decrease in blood PCO_2 would suggest a decrease in metabolic rate to less than 50% of the presubmergence rate. A similar decrease in metabolic rate had earlier been found to occur in the frog, *Rana temporaria,* while submerged in aerated water for three months at 3°C (Donohoe et al., 1998). These investigators used both direct measurement of aquatic O_2 uptake and the indirect evidence of decreased blood PCO_2 to reach their conclusion. For both the softshell (turtle) and the frog, the reduction in metabolic rate has the distinct adaptive advantage of sparing the overwintering animal's energy reserves and enhancing its chances for survival and reproductive success in the spring.

Two other species of turtle whose ranges also extend into northern latitudes are not as proficient at obtaining all of the oxygen they need from the water. The painted turtle, *Chrysemys picta,* and the snapping turtle, *Chelydra serpentina,* both show evidence of anaerobic metabolism when they are experimentally submerged in aerated water at 3°C (Jackson, Ramsey, Paulson, Crocker, and Ultsch, 2000; Reese et al., 2002). Unlike the map, softshell, and musk turtles, lactate concentrations in the blood of the painted turtle and snapping turtle increase significantly during submergence. This indicates that they cannot fully meet their oxygen demands by aquatic respiration and must recruit anaerobic metabolism to satisfy their energy needs. Because it is always desirable to confirm, when possible, the reliability of laboratory studies for describing natural behavior, we again launched a field study, this time on eastern painted turtles in a pond in Rhode Island. Gordon Ultsch was again the main instigator and Carlos Crocker once again braved the elements to carry out this project (Crocker, Feldman, Ultsch, and Jackson, 2000), with regular assistance during his trips to the pond from undergraduate student Rachel Feldman. The details of this study differed somewhat from the Vermont study. Turtles were collected from the pond's resident population in the early fall and were brought to our laboratory at Brown University, where they were fed earthworms and fitted with radio transmitters, each with a distinct signaling frequency. In October, we released the turtles back into their home pond and monitored their locations periodically by homing in on the transmitted signals with an antenna and a radio receiver. Beginning in December, Carlos and Rachel began collecting turtles on a monthly basis and analyzing blood samples, as described earlier with the map turtles. The turtles situated themselves in sites that were relatively shallow, so after opening a hole in the ice, Carlos could simply reach down and retrieve them. Blood data for these painted turtles differed from the map turtles. Blood lactate concentration was significantly elevated approximately twenty to thirty times the control level, confirming what had earlier been observed on turtles in the laboratory that were undergoing simulated hibernation. These results indicated that oxygen uptake from the water was inadequate to fully support aerobic metabolism, although the water oxygen partial pressure was high at each sampling period, close to the ambient level. Another similar field study that was carried out in British Columbia of naturally overwintering western painted turtles reported similarly elevated lactate levels (St. Clair and Gregory, 1990).

I actually participated in one of these winter excursions to the Rhode Island pond. The surface was frozen the day I was there but was covered with slushy water; I literally got "cold feet" in response to this experience and resolved to confine my future research activities to the heated confines of the laboratory. In addition, I experienced a brief moment of panic when I feared I had lost Carlos. He was walking toward the far side of the pond when the ice, not nearly as thick as in the Lamoille River, suddenly gave way under him and he sank into the water. I was standing a safe distance away but had neither a rope to throw to him nor a good idea how I could mount a rescue. Fortunately, Carlos did not sink out of sight and was able to wade to an area of thicker ice and haul himself out of the water. I breathed a sigh of relief as my heart continued to pound while the unflappable Carlos remained remarkably calm and unperturbed. Perhaps not as unflappable as he seemed, though, since Carlos's current research interests have led him to the Mohave Desert of California, about as far as you can go in the United States to escape from ice-covered New England waters.

Two further aspects concerning the painted turtle acquisition of O_2 from the water are worth discussing. First, in our laboratory study of this species, blood lactate increased as already described, but the increase occurred early in the submergence period, and by the twenty-fifth day lactate concentration had reached its peak and remained essentially unchanged for the remainder of the 125-day period underwater (Jackson, Crocker, and Ultsch, 2000). The stability of lactate indicates that for most of the time the painted turtle was submerged and not breathing, it was actually acquiring sufficient O_2 from the water to satisfy the O_2 requirements of its tissues. A reasonable interpretation of this observation is that the metabolic rate of the turtle fell sufficiently by the twenty-fifth day of submergence to permit the animal's limited skin-diffusing capacity to supply all of its O_2 needs. Alternatively, the turtle may have increased its diffusing capacity somehow, but this seems less likely.

The second aspect concerns the pathway that O_2 takes to enter the turtle's body. As noted earlier with regard to the map turtle, possible routes of entry include the skin, the buccal cavity, and the cloaca or cloacal bursae (see Chapter 2). Both of these latter pathways have been documented in other turtles, but each requires active energy-requiring pumping, whereas simple diffusion only requires that blood be supplied to the skin. In a separate study on painted turtles that were submerged in aerated water at 10°C, I, along with undergraduate students Rachel Feldman and Elizabeth Rauer and postdoctoral student Scott Reese, obtained evidence showing that in this species uptake of O_2 from the water occurs almost entirely by diffusion through the skin (Jackson et al., 2004). When we prevented a turtle from employing either buccal or cloacal pumping, we observed no effect on O_2 uptake; however, we cannot say for certain whether this is also the case for other species.

Do Sea Turtles Hibernate?

I discussed in Chapter 4 several trips that Henry Prange and I took to Tortuguero, Costa Rica, to study sea turtles. Henry and I were stimulated to take one further trip south to study sea turtles. During the unusually cold winter of 1978 immature loggerhead turtles weighing 18–64 kg were dredged up in surprisingly large numbers in mid-January from the bottom of the Cape Canaveral Channel by shrimp trawlers. When Archie Carr learned of this, he himself organized trawling runs during February and March of that year (Carr et al., 1980). During the February run some fifty-six turtles were collected, almost all immature or subadult loggerheads. Once on deck these turtles were observed to be very torpid and covered with black mud, which suggested that they had been buried for a long time. Their deep body temperatures were 13°–15°C, higher than the overlying water (11°C) but similar to the bottom mud. Although not stated explicitly, the implication was that these turtles survived. A month later, one hundred turtles were collected, this time in better condition and with body temperatures at about 19°C, the same temperature as the water. The conclusion was that these latter turtles were emerging from winter hibernation and that the February turtles were still in their overwintering hibernation state. This fascinating paper did not address the remarkable possibility that these relatively large reptiles remained submerged for perhaps weeks in cold water without surfacing to breathe. Our desire to study the physiology of these turtles was frustrated by the fact that the following winter was milder and no turtles were dredged up.

Another report suggesting hibernation-like behavior in sea turtles had been published a couple of years earlier. Scientists became aware of a population of green turtles in the Gulf of California being harvested by local fisherman from Mexico during the colder months of the year. These turtles rested on the seafloor in a torpid state and were easily collected, either with a harpoon or by simply carrying them up to the surface and loading them into waiting boats. In the paper reporting this finding, Felger et al. (1976) suggested that these turtles had been submerged in this state for one to three months. The authors also pointed out the vulnerability of the turtles and the danger that overharvesting could threaten the survival of this population.

Both of these reports of hibernating turtles strongly suggested that marine turtles could remain submerged for perhaps weeks at a time at temperatures that in each case were in the vicinity of 15°C or slightly below. For me this was an astounding assertion, because these turtles would not only be submerged for long periods but would almost certainly be surviving with little or no oxygen. Unlike the freshwater turtles described earlier, these are larger animals with smaller surface-to-volume ratios, and their higher body temperature would increase O_2 demand well above what could be supplied by diffusion through their thick skin. They would therefore have to rely largely or even entirely on anaerobic metabolism, because any oxygen they had on board at the start of their submergence would be exhausted in several hours at most. I later discuss (Chapter 6) the remarkable anaerobic capability of the freshwater painted turtle, to our knowledge the champion turtle at surviving without oxygen. But the highly adapted painted turtle can only survive a few days without oxygen at 15°C (Herbert and Jackson, 1985a). Obviously we must look more critically at these anecdotal claims concerning sea turtles.

The first major uncertainty is that the actual duration of continuous submergence in each case was unknown. I consider it likely that the duration was in fact only hours, rather than days or weeks. The second uncertainty is whether these turtles, particularly the Cape Canaveral loggerheads, were in a cold-stunned state and unlikely to ever recover and make it to the surface on their own.

Further doubt is cast on the possibility of long-term underwater hibernation in sea turtles by recent studies directly testing hibernating behavior in several species. Moon et al. (1997) slowly acclimated captive immature green and Kemp's ridley sea turtles to a low temperature by gradually dropping the tank temperature from 26°C to 11°C over a period of several weeks. They recorded breathing frequency and observed that even at the lowest temperature the turtles never remained submerged for more than approximately three hours. They never observed the long periods of breath holding suggested by the earlier field observations. Some years earlier, David Kraus, an undergraduate student in my lab, and I measured the breathing of immature green turtles at 15°, 25°, and 35°C and never saw a cessation of breathing for more than a few minutes, even at the lowest temperature (Kraus and Jackson, 1980). Our acclimation time was only one day at each temperature, however, and our turtles were restrained, so this was not as good a test of the hibernation possibility as the study by Moon and associates (1997) was.

An even better test, though, was a recent field study by Hochscheid et al. (2007) on the natural diving behavior of loggerhead turtles in the Mediterranean area. These investigators installed data loggers that could capture time, depth, and temperature information and transmit these data to a satellite when the turtles surfaced to breathe. The Advanced Research and Global Observation Satellite (ARGOS) that was employed also identified the position of each turtle throughout the sampling period. Complete records were obtained for nine turtles throughout the winter at temperatures down to 12°C, and the results indicated that breathing episodes occurred regularly and that dive durations never exceeded eight hours. The authors believe that these dives were within the aerobic dive limit of these turtles because the surface breathing durations were too short to clear substantial lactic acid from the system. In a similar study by Broderick et al. (2007) on both loggerhead and green turtles, the longest overwintering breath hold recorded for a green turtle was just over five hours and for a loggerhead turtle was over 10 hours, the latter, according to the authors, the longest breath hold ever recorded for a marine vertebrate. It is likely, based on these studies, that marine turtles spend some time at a low temperature quietly resting on the bottom, but never for periods of days or weeks, as suggested by the observations of Felger et al. (1976) and Carr et al (1980). Nevertheless, I think the latter observations are sufficiently intriguing for the subject to remain open to potentially new information.

I was indirectly involved in one other study of the diving physiology of sea turtles that has some relevance to these turtles' survival and management (Schwartz, 2001). I served as the off-campus advisor for Malia Schwartz, who was a graduate student at the University of Rhode Island. Her interest was in understanding the responses of sea turtles that become trapped in shrimp nets and often drown. The turtles' vulnerability to being held underwater without breathing is already anecdotal evidence that these animals are not able to endure prolonged submergence. Malia traveled to Grand Cayman Island, where she studied captive green turtles at the Cayman Turtle Farm. Her subjects were immature animals weighing 10–20 kg. To simulate entrapment she forcibly submerged the turtles for thirty minutes at 25°C and observed that this was highly stressful to them and left them in a very weakened state with elevated blood lactate levels. The turtles recovered from this experience but clearly could not have survived much longer without breathing.

Because of the stress factor, this experiment does not reveal much about the normal voluntary diving behavior of green turtles, but it does reveal why entrapment in shrimp nets is a serious threat to these animals. Their struggles to escape drastically shorten their survival time and can explain why so many sea turtles succumb.

Concluding Comments

From the foregoing, it is clear that overwintering freshwater turtles can remain submerged for many months without breathing with, at most, modest disturbances in their physiological homeostasis if the water has high oxygen partial pressure. Unlike humans, other mammals that are natural divers, or even marine turtles, the freshwater turtles I have described do not rely entirely on the O_2 they have on board at the start of the dive. They can slowly acquire the O_2 they need directly from the water and supply the modest needs of their cold tissues, probably largely by diffusion of

O_2 through their skin. They also lose the CO_2 they produce by the same route.

The situation for a hibernating turtle could become perilous, however, because oxygen levels in many bodies of water, such as stagnant ponds, can become severely depleted when oxygen use is high by resident organisms and/or when oxygen-producing photosynthesis is blocked by a heavy snow cover. If a turtle finds itself in such an environment, it may not have adequate O_2 available to support aerobic metabolism and will have to rely more on anaerobic metabolism. This can drastically shorten the time it can survive without breathing. Turtles that have the most effective aquatic gas exchange, the map, softshell, and musk, are handicapped most in water without oxygen. For these turtles, finding a location to hibernate that is reliably oxygenated, such as the flowing river in Vermont, is a crucial consideration for surviving the winter. In contrast, the painted turtle and the snapping turtle, species less adept at aquatic respiration, are remarkably tolerant to the absence of oxygen, and this provides them with more options regarding where they can hibernate. Surviving for long winter months at a low temperature with ample O_2 available in the water is already a challenge for a turtle, so how can the turtle survive with little or no O_2?

6

LIVING WITHOUT OXYGEN

At regular intervals throughout the winter, we entered the cold room down the hall from my laboratory, drew blood samples from the submerged turtles, and evaluated the turtles' condition. Week after week and month after month, my colleague, Gordon Ultsch, and I were amazed to find turtles that were still responsive. These turtles, western painted turtles *(Chrysemys picta bellii),* were in very cold water (3°C) that was continuously bubbled with pure nitrogen gas; in other words, the turtles were anoxic, meaning they had no oxygen. They were unable to breathe, just as would be the case if they were trapped under the ice of a frozen pond, but unlike the overwintering turtles described in Chapter 5, these turtles had no dissolved O_2 in the water from which to draw.

Humans and most other vertebrates can survive for only a few minutes without O_2. The capacity to survive for such a long period is without question the turtle's most remarkable physiological trait. It is also a trait that is crucial to help the turtle survive cold winters, because the painted turtle often hibernates either in anoxic mud or in ponds that become depleted of O_2 (Ultsch, 2006). Discovering how a turtle can live without O_2 has been a major focus of my research for many years, and I describe in this chapter some of what I and many others who are equally fascinated by this remarkable phenomenon have learned.

We have known for many years that turtles have an unusual tolerance to anoxia (e.g., Johlin and Moreland, 1933). A particularly noteworthy experiment was performed by Daniel Belkin at the University of Florida. He collected some four hundred individual reptiles, including turtles, lizards, snakes, and crocodilians, and determined how well each animal fared when exposed to an oxygen-free environment. He placed the animals, one by one, into a glass chamber filled with nitrogen gas and measured the time that elapsed between the first breath of nitrogen and the last breath before the animal stopped breathing for a set period. He then re-suscitated the animals. His results revealed a striking difference among the different reptiles. The turtles, with the exception of the marine turtles, lasted, on average, approximately twelve hours, far longer than the other reptiles that reached their end point in approximately forty-five minutes. The marine turtles were intermediate at two hours. Belkin concluded that turtles, for reasons that were uncertain, are considerably more tolerant of the absence of oxygen than other reptiles (Belkin, 1963).

Not long after Belkin's pioneering study Eugene Robin and his colleagues, who worked at the Mount Desert Island Biological Laboratory in Maine, reported that turtles of the genus *Pseudemys* (species not named) could survive for a day or more either held underwater or while breathing nitrogen gas (Robin et al., 1964). They also collected blood samples and measured large increases in blood lactate concentration that were associated with increases in the acidity of the blood samples.

These pioneering studies, all conducted at room temperature, demonstrated that freshwater turtles are extraordinary among the vertebrates in their ability to survive for long periods without oxygen. What these studies did not reveal, however, was how the turtles do this and what they have that their fellow vertebrates lack. However, these studies did stimulate further investigations of the subject by many researchers, including me.

Around the time the studies by Belkin and Robin were published, I had the good fortune of doing postdoctoral work in the laboratory of Knut Schmidt-Nielsen at Duke University. Professor Schmidt-Nielsen, who was one of the foremost comparative physiologists of the twentieth century, had suggested a project that involved measuring the heat production of the toadfish, *Opsanus tau,* using the method of direct calorimetry. Direct calorimetry measures the amount of heat generated by an animal's metabolism. The toadfish, a marine fish that inhabits waters off the East Coast of the United States, was thought at the time to be an oxygen conformer. An oxygen conformer is an animal whose metabolic consumption of oxygen varies directly with how much oxygen is in the water it breathes. If the surrounding oxygen goes down, then the rate at which the fish consumes O_2 also goes down, unlike a so-called oxygen regulator that maintains a constant rate of consumption until ambient oxygen reaches a critical low level. My goal was to determine whether the toadfish's total metabolic rate (a quantity only measurable as heat production) remained stable even as its oxygen consumption fell in response to reduced oxygen in the water. The hypothesis that Knut wanted me to test was that anaerobic metabolism (metabolism that does not require oxygen) would increase to compensate for the lost oxygen consumption and thereby keep total metabolism unchanged. To do this study I would have to measure total heat production using a method known as direct calorimetry.

The opportunity to use direct calorimetry appealed to me because I inherited a legacy based on this technique. My PhD advisor, Ted Hammel, employed calorimetry in his studies of heat regulation in mammals. Hammel's mentor, James Hardy, used the method to study human heat exchange at Cornell Medical School, and Hardy's mentor, Eugene F. Dubois, also used the technique as did his teacher, Graham Lusk. From Lusk, the scientific family tree can be traced back through Carl von Voit, Baron Justus von Leibig, Joseph Louis Gay-Lussac, and Pierre Laplace, all the way to Antoine Lavoisier, the French chemist who, with the assistance of Laplace, built the very first animal calorimeter in the latter part of the eighteenth century. I was therefore the proud descendant of this long line of distinguished scientists, many of whom had utilized the very technique I was now setting up for my experiments.

Unfortunately, after assembling the apparatus to attempt the toadfish study, it became painfully obvious to me that to simultaneously measure oxygen consumption and heat production with the approach I was using was technically unfeasible. So there I was with my lab bench all dressed up and nowhere to go. This was an embarrassing situation for a young researcher trying to get started in the scientific world, particularly one who was attempting to continue a rich scientific tradition. At this critical moment, I read the 1964 publication by Robin and colleagues (previously cited) and a lightbulb lit up in my brain. I realized that I could substitute turtles for toadfish and carry on in a different direction. Now my goal would be to learn how the total metabolic rate of a turtle, as measured by its heat production, changes when the turtle becomes anoxic. Knut graciously gave his blessing to this change in direction, and thus began my career studying turtles.

Implications of Anoxia Tolerance in Turtles

An inescapable inference from the observations by Belkin (1963) and Robin et al. (1964) is that a turtle can support all of its vital processes without any oxygen whatsoever for long periods. This is radically different from what is possible in any mammal or bird, because we and our fellow warm-blooded animals all require a continuous supply of oxygen to sustain our brains and hearts. Breath-hold duration, whether in a human, an elephant seal, or a penguin, is limited by how long the oxygen lasts. Although some naturally diving mammals and birds can endure very low blood levels of oxygen (Meir et al., 2009; Ponganis et al., 2007), none can survive, as far as is known, when all of their oxygen is gone. This means that a turtle is fundamentally different because its central nervous system and its heart, as well as the rest of its body, can continue to function at summer temperatures for many hours and at winter temperatures for many weeks with no oxygen available. All essential cellular processes can be supported by the anaerobic (i.e., zero oxygen) production of energy.

In all animals, including turtles, metabolic pathways capture energy contained in the chemical bonds of fuel molecules, such as fat and carbohydrates, and store this energy in a high-energy chemical bond of the molecule adenosine triphosphate (ATP). The release of the energy stored in this chemical bond provides the immediate source of energy required by cells for their living processes. In an anoxic turtle, all of this ATP must be produced by anaerobic metabolism.

To survive without oxygen, a turtle, particularly when enduring many winter weeks in this state, must not only maintain a viable level of ATP but must deal with two additional threats to its survival: exhaustion of its fuel supply and severe acidosis. The fuel supply problem is a straightforward one. At low temperatures near freezing, a turtle does not feed and must therefore rely totally on substances already stored within its body that it can metabolize for energy. For most starving animals the fuel substance of choice is fat, because it is readily stored in ample amounts and is a high-energy source. Fat, however, requires oxygen for its metabolism and cannot be used by an anoxic animal. Instead, an anoxic turtle must use stored carbohydrate, which exists in the body partly as a simple sugar, such as glucose, but primarily as the complex molecule glycogen. Glycogen is a polymer composed of many glucose molecules and is stored chiefly in the liver and in skeletal muscle. An adequate supply of glycogen must be on board, because when glycogen is depleted no further ATP can be produced. The complicating factor for an anoxic animal is that anaerobic metabolism is quite inefficient in its use of carbohydrate because it captures only a small fraction of the chemical energy available in each glucose molecule. Only two ATP molecules are produced per glucose molecule in the absence of oxygen (three ATP molecules if glycogen is the immediate source of glucose) compared to up to thirty-six ATP molecules per glucose molecule when the presence of O_2 permits the complete oxidation of glucose to CO_2 and water. What this means is that to generate the same amount of ATP from glycogen, an anoxic animal must consume glycogen up to twelve times faster than an aerobic animal.

The second major threat facing an anoxic turtle is that this same profligate anaerobic pathway dead ends with the production of lactic acid. Most readers have some experience with this substance because our muscles produce lactic acid when we engage in exhaustive exercise. Lactic acid is a relatively strong acid that dissociates at physiological pH almost completely (~99.9%) to a proton (H^+) and a lactate ion, so most of the lactic acid in the body exists as lactate, a negative ion, or anion. To survive while anaerobic, a turtle must somehow neutralize the lactic acid in order to prevent a severe fall in body fluid pH. The duration of anoxia, therefore, may eventually be limited by an excessive buildup of lactate in the body and a fatal acidosis. The faster the turtle produces lactic acid, the sooner the dire consequences of intolerable acidosis will occur.

It is now possible to appreciate the importance of an anoxic turtle's metabolic rate, the variable I was setting up to measure with my rejuvenated calorimeter. Both the rapidity at which glycogen reserves are depleted and the rate at which lactic acidosis reaches dangerous levels depend directly on the intensity of the anoxic turtle's anaerobic metabolism. If it is possible for this intensity to be very low, so that glycogen loss and lactic acid production occur very slowly, then the duration of anoxic survival can be much longer.

Adaptations to Anoxia: Metabolic Depression

A turtle's heat production does indeed fall dramatically when the animal becomes anoxic. Both the study I conducted at Duke University when I switched from toadfish to turtles (Jackson and Schmidt-Nielsen, 1966) and the follow-up study I conducted at the University of Pennsylvania (Jackson, 1968) showed that a turtle's heat production decreases by about 80%–90% from the value measured when the turtle breathes air at the same temperature. The species of turtle I studied in each case was the red-eared slider, *Trachemys scripta elegans*. The second study also revealed that metabolic rate did not fall until the turtle's oxygen stores were almost gone and heat production reached its lowest level when the turtle's metabolism was completely anaerobic. These studies provided the first evidence that turning down the rate of metabolism is a key mechanism that allows a turtle to prolong the time it can survive without oxygen. However, I performed both of these studies at room temperature (24°C), a convenient temperature for laboratory investigation and a feasible temperature for measuring heat production, but not a temperature at which a turtle in nature is likely to become anoxic. When freshwater turtles are free to breathe, as they normally are at temperatures above freezing, they avoid anoxia by rising periodically to the surface to breathe. On the other hand, if submerged during the winter in an ice-covered pond with no dissolved oxygen, then a turtle cannot avoid anoxia.

This brings us back to the study of anoxia in turtles at cold temperature, described at the beginning of this chapter. This study, a collaborative effort with Gordon Ultsch of the University of Alabama, was carried out in my laboratory at Brown University. We performed a study that simulated hibernation in the western painted turtle, *Chrysemys picta bellii*. We chose this species because its habitat extends into Wisconsin, Minnesota, and neighboring parts of southern Canada, where ponds can be ice-covered for months. The experiments we conducted on these animals were ecologically relevant and resembled the challenges that these turtles face each winter in their natural habitat.

My department had a large walk-in cold room that we set at 3°C, a temperature that a turtle sitting at the bottom of a frozen pond might encounter. Before the study began, we surgically placed small tubes (catheters) into an artery of each turtle and then slowly cooled down the turtles, over a two-week period, to 3°C. We placed groups of turtles in several water-filled tanks in this room, but for this discussion, the most important tank was the one we bubbled continuously with nitrogen gas to keep the water close to zero O_2. The turtles in this tank were anoxic. As noted earlier, we periodically entered the cold room, removed the sheet of Plexiglas sealing the top of the tank, determined the condition of the turtles, and, with our benumbed hands immersed in the 3°C water, collected blood samples from each of the turtles. This experiment lasted far longer than we expected, and our results revealed that these remarkable animals can stay alive during an entire winter with no oxygen (Ultsch and Jackson, 1982).

The sluggish state of these cold, anoxic turtles made it clear that their rate of heat production was very low—but how low? Because direct calorimetry, the method I used at 24°C, was not technically feasible at a very cold temperature, an indirect estimate had to be made based on the accumulation of lactate in the turtle's body. Data collected by graduate student Christine Herbert revealed that metabolic heat production during anoxia decreased by about 90% from the already very low aerobic rate at the same temperature (Herbert and Jackson, 1985b). This decrease due to anoxia was almost the same, percentage wise, as I had earlier observed at 24°C using direct calorimetry (Figure 6.1). However, at 3°C, the absolute rate of metabolic heat production was far lower. Our calculations revealed that the turtles were generating heat at the feeble rate of 0.009 calories per kg of body weight per minute. A calorie is the amount of heat required to raise the temperature of 1 g of water 1°C and is equal to 4.18 joules.

Because I recognize that this numerical description of metabolic heat production by a cold, anoxic turtle may not be particularly informative for every reader, let me give a graphic example of what it means. First, consider your own metabolism. If you are sitting quietly and not performing any external work, then all of your metabolic energy in the steady state ends up as heat. Suppose, for argument's sake, that all of this heat remains within your body rather than being lost to the surroundings, as it normally is. Your body temperature will inexorably rise and reach a lethal level of about 42°–43°C in roughly four to five hours. Now consider an even more extreme example of heat storage, one that actually occurs. The cheetah is the fastest land animal on earth and has been clocked at about 110 km/hour (68 mph). Richard Taylor and Victoria Rowntree trained a cheetah to run on a treadmill at their lab at Harvard University and measured the cheetah's body temperature as it ran. They discovered that a cheetah stores most of the heat it produces during running; consequently, its body temperature rises rapidly (Taylor and Rowntree, 1973). These investigators also observed that when the body temperature of a cheetah rose to about 40.5°–41.0°C from its normal temperature of 37°C, the cheetah refused to run any farther. In the authors' words, "They would simply turn over with their feet in the air and slide on the tread surface" (850). Based on the data they obtained at experimentally feasible treadmill speeds, the authors extrapolated what the thermal consequences would be for a cheetah running at its top speed. Their conclusion was that the body temperature of a cheetah hurtling across the African savannah at 110 km/hour would rise at the astounding rate of 1.6°C per minute and would reach a critically high body temperature by the time it had traveled about 1 km. Significantly, this is the

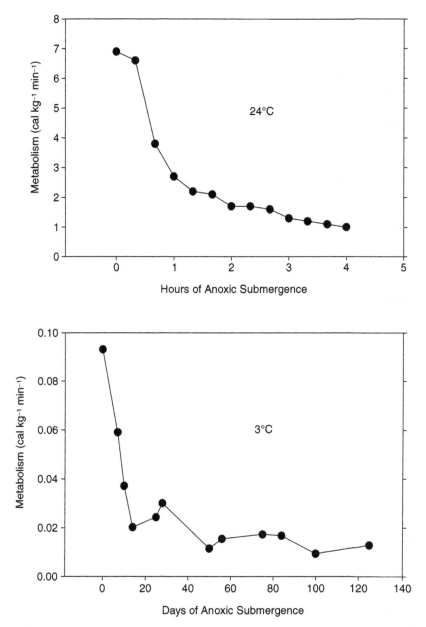

Figure 6.1 Metabolic heat production of anoxic turtles, red-eared sliders, at 24°C, measured by direct calorimetry, and western painted turtles, at 3°C, estimated from lactate buildup in body. Note the differences in metabolic scales (from Jackson, 2002).

approximate distance a cheetah chases its prey in the wild before abandoning the hunt, which suggested to the authors that rising body temperature may limit how far and how long a cheetah can run.

Now let us return to the turtle for a similar calculation. How fast would the body temperature of a cold, anoxic turtle rise if the turtle were to store all of its metabolically produced heat? The amazing answer, based on the estimated heat production of 0.009 calories $(kg\ min)^{-1}$, is that it would take about two and a half months for the turtle's temperature to increase by 1.0°C. This is only 0.00001°C per minute compared to 1.6°C per minute for the sprinting cheetah. In other words, each gram of a cheetah sprinting at its top speed generates heat 160,000 times faster than each gram of a cold, anoxic turtle resting quietly on the bottom of a frozen pond. The cheetah is a roaring bonfire compared to the barely burning ember of an anoxic turtle (Figure 6.2). Even our modest heat production at rest, which is about the same as the heat given off by a 100-watt incandescent lightbulb, would raise our body temperature more than two thousand times as fast as the turtle's if no heat were lost from the body.

It is also noteworthy that the metabolic rates of both a sprinting cheetah and a cold, anoxic turtle are powered almost totally by the same anaerobic metabolic pathway. We are left, therefore, with the remarkable conclusion that this rather inefficient metabolic pathway that yields only two ATP molecules per glucose molecule can support what are close to the highest and the lowest rates of metabolism that can be found among all the vertebrates. In both the galloping, overheating cheetah and the cold, torpid turtle, the length of time that the anaerobic activity continues is self-limiting, in the cheetah perhaps by the explosive rise in body temperature, and in the turtle, and perhaps also in the cheetah, by glycogen depletion and acidosis.

The profound depression of function exhibited by the cold, anoxic turtle is also evident in the turtle's heart function. When Christine and I conducted the submergence study mentioned earlier, the turtles had also been surgically equipped with arterial catheters for collection of blood samples. Toward the end of a three-month submergence period at 3°C, we connected the catheter of one of these anoxic turtles to a pressure transducer and a polygraph recorder to provide a trace of the turtle's arterial blood pressure. After making all of the connections, we watched the polygraph pen moving on the chart and, to our dismay, the trace of the pen appeared to be a flat line near the bottom of the chart. We waited. Several minutes passed and still we saw no apparent change in the recording. We

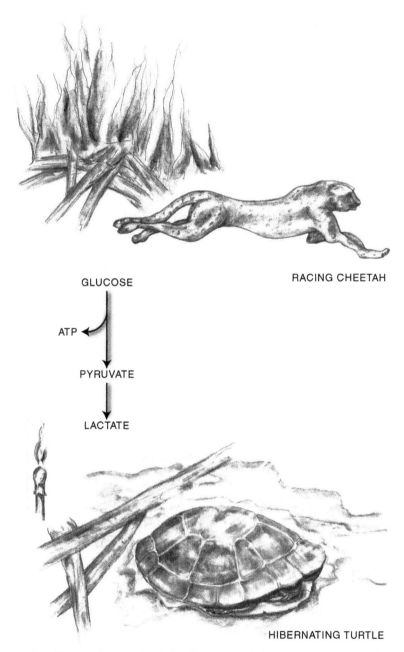

GLUCOSE

ATP

PYRUVATE

LACTATE

RACING CHEETAH

HIBERNATING TURTLE

Figure 6.2   The roaring metabolic bonfire of a sprinting cheetah and the faint ember of a cold, anoxic turtle.

were starting to fear that the turtle had no blood pressure when slowly, very slowly, the pen began to move upward. It reached its highest level and then even more slowly began to descend toward the baseline. The turtle's heart had finally produced a beat. We continued to record the pressure, and sure enough additional beats continued to occur at long intervals, the longest interval between beats being an astounding ten minutes! In a human heart at rest, the interval between beats is normally one second or less. At summer temperatures, a turtle's heart also beats much faster. In Figure 6.3 the slow heartbeat of a cold, anoxic turtle is compared to that of a warm turtle breathing air.

The anoxic turtle's peak blood pressure was also low, averaging only about a tenth of our peak blood pressure. When the three-month submergence period was over, this turtle and others undergoing a similar procedure were again able to breathe air. Once breathing resumed, their heart

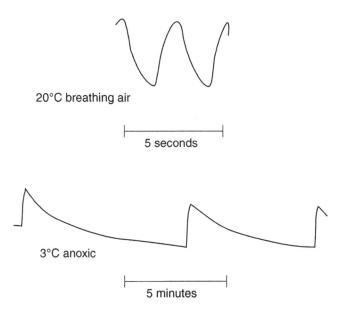

Figure 6.3   Blood pressure recording from a red-eared slider breathing air at 20°C (from Shelton and Burggren, 1976) and from a painted turtle submerged with no oxygen at 3°C. Each deflection of the line represents a heartbeat. Note the difference in time scales for two records. The heart rate at 20°C is 168 times faster than at 3°C. Adapted with permission from the *Journal of Experimental Biology* 64: 323–343.

rate and blood pressure increased to a normal level for this temperature, a heart rate of about two beats per minute and a blood pressure twice the anoxic value.

Metabolic depression is not unique to anoxic turtles. It occurs in response to a variety of stressors in addition to anoxia, including extreme temperature, desiccation, high salinity, and food deprivation. This response to stress also occurs in a variety of organisms representing nearly all of the major animal phyla (Guppy and Withers, 1999). Organisms lower their metabolism to sustain themselves through periods of hardship. The cellular mechanisms involved in metabolic depression are beginning to be unraveled, and the results could potentially have major clinical and practical importance to humans. If, for example, the metabolic rate of a trauma victim could be lowered during transport to a medical center, then the probability of a favorable outcome could be greatly enhanced. Or, suppose patients who are now made hypothermic for certain surgical procedures could instead be depressed metabolically at their normal body temperature by some procedure that simulates what occurs in anoxic turtles. Even the familiar sci-fi fantasy of space travel in a hypometabolic state is possible in principle if the secrets of animal metabolic depression could be unraveled.

Metabolic Depression at the Cellular Level

Studies of anoxic turtles at the cellular level are helping us understand the basic mechanisms of depressed metabolism. Important work has been done by Peter Hochachka and his colleagues, who studied responses to anoxia of isolated liver cells (hepatocytes) taken from painted turtles. In one of these studies (Buck et al., 1993) heat production, measured directly using a microcalorimeter, fell by 76% when the cells were made anoxic, a fall in heat production very similar to what I had observed earlier on whole animals. In further studies (Hochachka et al., 1996) these investigators were able to show that specific energy-requiring processes in hepatocytes were suppressed during anoxia by anywhere from 70%–100% of the aerobic level. The energy-requiring processes they studied included sodium ion transport and protein synthesis.

Pumping sodium ion (Na^+) out of cells is of fundamental importance because it maintains the chemical and electrical gradients across cell membranes that are necessary for establishing a cell's resting membrane potential, producing the action potentials of nerve and muscle, and supplying energy to transport many other substances in and out of cells. The energy required to pump Na^+ out of a cell in exchange for K^+ is a major consumer of ATP in all cells. Given its crucial importance, how then can the metabolic energy required to accomplish this vital function be drastically reduced in an anoxic turtle? The answer is that the energy needed to pump sodium out is determined by how much sodium leaks in. Perhaps an analogy will help here. Imagine you are in a rowboat that has several holes in the bottom through which water is pouring in. To stay afloat, you must bail out the water as fast as it is coming in. If you could somehow plug up some of the holes, then less water would enter the boat, and you could reduce your bailing effort yet still maintain the same water level in the boat. Similarly, the cost of sodium pumping is less in anoxic turtle cells because their cells become less leaky to sodium. Leakage occurs through water-filled pores called ion channels, which alternate between an open and a closed state. During anoxia, these channels spend more time in the closed state, which reduces sodium entry into cells. Less sodium leakage means less sodium that must be pumped out to maintain the same ion gradients, thereby the cellular energy expenditure is turned down. Hochachka called this reduction of ion leakage "channel arrest."

Cellular processes such as the Na^+ pump and protein synthesis are ATP consuming processes. They use the energy stored in the ATP molecular bond to pay for moving an ion uphill against its gradient or for synthesizing a new macromolecule. The suppression of these and other energy-demanding cellular activities in the anoxic turtle lowers the overall use of ATP by some 80%. However, in order for the cell to remain in energy balance, the rate of production of ATP must also fall by some 80%. In other words, energy supply and energy demand must be coordinated. The pathway operating in the anoxic turtle that produces ATP is the classic pathway called glycolysis, which converts glucose or glycogen to pyruvate and then to lactate. When the turtle has access to oxygen, the pyruvate is further metabolized to CO_2 and H_2O via what are called the Krebs cycle and the electron transport system and, as discussed earlier, generates much more ATP in the process. What if no O_2 is present? The classical biochemical answer is that anaerobic metabolism is strongly activated in order to keep ATP production relatively constant, despite the cost of greater fuel utilization. This response to oxygen lack is known as the Pasteur effect. This response is not, however, what occurs in turtles, or in their isolated cells, or in other animals that are tolerant of anoxia. Instead, the rate of metabolism by this pathway changes little, or it may even slow down, thereby greatly reducing the rate of ATP production. This has been called the "reverse Pasteur effect," and it helps match the production of ATP to the breakdown of ATP. Kenneth Storey (1996) has studied the control of anaerobic metabolism in the tissues of turtles and other organisms tolerant to oxygen lack and has identified several regulatory effects on enzymes of the anaerobic pathway that contribute to down regulation. Exactly how these two processes, reduction of ATP use and reduction of ATP production, are matched to maintain the cells' energy balance is still not well understood, although the biochemical mechanisms involved are beginning to be discovered (Staples and Buck, 2009).

A particularly remarkable aspect of the turtle's tolerance to anoxia is the continuing functional integrity of its brain and nervous system in the absence of oxygen. The fact that the turtle brain can survive with no oxygen is stunningly different from most other vertebrates and has attracted considerable research interest. The turtle brain's response to anoxia is a coordinated one, involving drastically reduced neuronal activity and metabolism, maintenance of ATP levels and ionic gradients, and a defense of structural and functional integrity so that normal activity can resume when oxygen is restored (Lutz and Milton, 2004). A fundamental and crucial response of the turtle brain is channel arrest, the reduction of the passive flux of ions such as Na^+, K^+, and Ca^{2+} across cell membranes through ion channels. Channel arrest reduces the work required to maintain ion balance within cells and also avoids the catastrophic influx of calcium ions that triggers destructive cellular events in the anoxic mammalian brain (Bickler and Buck, 1998). In addition, the massive release of excitatory neurotransmitters, such as glutamate, to neurotoxic levels contributing to anoxic death in mammalian brains does not occur in turtle brains. It is probably safe to say that during a turtle's long weeks of cold anoxia, its brain is not engaged in much creative activity, but this vital organ is intact, alive, and prepared to resume normal cognitive function come springtime.

Adaptations to Anoxia: Buffering the Lactic Acid

When Gordon Ultsch and I did our initial study of cold, anoxic turtles, we collected blood from these turtles at regular intervals over the course of their submergence, in some cases up to five months. Among the measurements we made were blood pH and lactate concentration of the blood plasma (the cell-free fluid of blood). As anoxic submergence proceeded, the animal's blood pH fell and plasma lactate went up, as expected, but lactate concentrations rose to unusually high values, in the extreme case to an astounding 200 mmol per liter. Again, this number may not be meaningful to many readers, but it may be helpful to compare it to our own physiology. In both a turtle and a human normal plasma lactate concentration is about 1 mmol per liter. If a human athlete exercises to exhaustion, then plasma lactate may rise to an extreme value of 20–25 mmol per liter. Yet some of our anoxic turtles had lactate concentrations up to ten times this high. Because these values were unprecedented in previous studies, we sought independent verification from our respected colleague, Peter Hochachka, at the University of British Columbia, who kindly repeated our measurements and confirmed that we were correct. Then our challenge was to understand how a turtle could survive such a massive acid load.

Clearly the anoxic turtle must find some way to neutralize acid by some buffering mechanism. The problem facing the turtle, and the problem we were trying to fathom, is that the amount of acid the anoxic turtle had generated vastly exceeded the amount of buffer normally present in the turtle's body fluids. The major buffer in the blood plasma and the rest of the extracellular fluid for neutralizing lactic acid is bicarbonate ion (HCO_3^-). This buffer is familiar to anyone who has taken baking soda to treat acid indigestion. Baking soda is sodium bicarbonate, and its action is to neutralize the acid in the stomach, which is hydrochloric acid (HCl). When the acid reacts with the bicarbonate, the products are CO_2 (the familiar burp that usually follows this treatment is CO_2 gas being expelled from the stomach) and H_2O (where the pesky H^+ ions now harmlessly reside). The reaction with lactic acid is similar and can be depicted chemically as follows:

$$H^+ + lactate^- + Na^+ + HCO_3^- \rightarrow Na^+ + Lactate^- + CO_2 + H_2O$$

Turtles have high concentrations of bicarbonate in their extracellular fluid compared to other vertebrates, nearly twice as much, for example, as we have, but even this unusually high concentration is not nearly adequate to buffer the enormous amount of acid that accumulates in the body during these long bouts of anoxia. Yet somehow the pH of the turtles fell only moderately during months of anoxia, so the turtles must have recruited additional buffering elsewhere. What is this additional buffer, and where does it come from?

We assumed that the additional buffer was also bicarbonate (or perhaps carbonate) and that it entered the blood either accompanied by a positive ion, such as sodium (Na^+), or in exchange for a negative ion, such as chloride (Cl^-). As in our stomachs, the bicarbonate would be converted to CO_2 and water. The turtles could not eliminate the CO_2 by burping, but the CO_2 could diffuse out of the body through the skin and be lost from the system. Because the CO_2 is lost, the evidence for the additional buffering would thus have to be an increase in Na^+, or some other cation, or a decrease in Cl^-, or some other anion. We therefore looked for changes in the ion composition of plasma to provide evidence for added buffer. The ions that we measured initially, besides lactate, were Na^+, K^+, Cl^-, and HCO_3^-. As expected, HCO_3^- concentration fell, plasma Cl^- also fell, and K^+, but not Na^+, increased. We still had a problem, however. An aqueous solution must satisfy the requirement for electrical neutrality; that is, the total concentration of positive charge, in our case ($Na^+ + K^+$), must equal the total concentration of negative charge, again in our data (lactate$^- + Cl^- + HCO_3^-$), but when we made this calculation we discovered we were missing a large amount of positive charge. Some major cation, or cations, was not being measured.

At this point, we enlisted help from colleagues to test for various cations that could potentially satisfy this large discrepancy in charge. We measured amino acids that have positive charge but found no significant changes. We measured ammonium ion (NH_4^+), but the concentration measured from the anoxic turtles was very low, so this possibility was ruled out. What could the missing cation be? The obvious choice that stared us in the face was calcium, Ca^{2+}, a strong divalent cation, but we found it hard to believe this was possible because calcium is very tightly regulated in vertebrate plasma. For example, in our bodies, normal calcium concentration is about 2.4 mmol per L; even a modest increase to 3 mmol per L can produce serious symptoms, and an increase to 4 mmol per L can produce coma or cardiac arrest. If calcium were the missing cation in our calculation, then its concentration would have to be much greater than this, and we doubted that a turtle could survive such a high level.

Despite this unlikelihood, we decided to test for calcium, and also for magnesium, another divalent cation normally present in plasma. A colleague in geological sciences at Brown, Professor R. K. Matthews, generously agreed to let us bring over some of our plasma samples and use his atomic absorption spectrophotometer, an instrument that can measure the total concentrations of both calcium and magnesium in a fluid sample. He also provided the services of his skilled assistant, R. A. Fifer. This instrument displayed its results on a chart recorder that produced a deflection when it detected a particular element, and the height of the peak, with suitable calibration, indicated the concentration of that element in the sample. The sensitivity was initially set to detect calcium at its usual plasma level.

When Fifer injected the sample, Gordon and I watched the recorder anxiously. Up, up, up the needle rose, and we nearly fell over as the pen's rapid climb caused it to pin against the top of the chart. The concentration of calcium was very high indeed. In order to get a peak that stayed on the chart and thereby determine the actual concentration, we had to dilute the samples manyfold. Using this approach, we analyzed a large number of plasma samples with this instrument and found that both calcium and magnesium rose steadily during anoxia, in parallel to lactate, from starting values near those cited for humans, up to twenty-five times as high for calcium and almost ten times as high for magnesium. The extraordinarily high values of these two cations proved to be the missing positive charges and almost perfectly satisfied the condition of electrical neutrality in the plasma samples.

The next uncertainty was the source of all the calcium and magnesium. Well, what do you see when you look at a turtle? You see its shell. What is the major element in the shell? Calcium. In fact, over 99.9% of the calcium a turtle has in its whole body is located in its shell and in the parts of its skeleton that are not part of the shell (its skull, it limb bones, and its pectoral and pelvic girdles). These structures also hold the bulk of the body's magnesium; therefore, the shell was the likely major source of the extra buffers. The only other possible source could be from within cells, but in a later study with Norbert Heisler at his lab in Germany, I found that calcium actually increases in concentration in muscle cells during anoxia (Jackson and Heisler, 1983).

As discussed in Chapter 1, the turtle's shell closely resembles the human skeleton in its composition. The primary mineral is calcium phosphate but, like our bone, the shell and skeleton of the turtle also contain other elements and molecules, some in rather large amounts. In the painted turtle these structures house over 90% of the magnesium, over 60% of the sodium, and over 95% of the CO_2 (mostly in the form of carbonate). Both phosphate and carbonate are buffer ions, which means that the shell and skeleton, which together comprise what I will call the turtle's bone, contain the bulk of the body's potential buffering capacity. Thus when we measured large increases in calcium and magnesium in the plasma of turtles after several months of anoxia at 3°C, we assumed that these elements must have come largely from the shell. We also assumed that they came out of bone, along with buffer anions—carbonate and/or phosphate.

Buffer Release from Shell

The possibility that bone was a source of buffering for the turtle was not a radical idea. The skeletons of mammals, including humans, release buffers in response to acidosis. Both experimental evidence and clinical evidence have also shown that calcium carbonate and/or calcium bicarbonate are the probable buffer compounds in mammals. Our results indicated that calcium and magnesium carbonates are also the major buffers released from the bones of anoxic turtles. However, our observations of the anoxic turtle were unusual because of the magnitude of its response and because the calcium (and magnesium) increased to very high concentrations in the blood. In mammals, blood calcium stays within very narrow limits, even during acidosis, because the calcium released from bone is excreted by the kidneys. In an anoxic turtle, the kidneys are practically nonfunctional (Warburton and Jackson, 1995), and so the calcium and magnesium accumulate in the blood. Because the carbonate that reaches the blood reacts with H^+ to form first bicarbonate and then CO_2, and because the CO_2 is lost to the water surrounding the turtle, the "smoking gun" for this buffering mechanism by turtle bone is the large rise in the concentrations of calcium and magnesium. The other potential buffer of bone, phosphate, does not appear to participate in this process.

Because of the loss of so much calcium from the turtle's shell, one would suppose that the shell would become significantly thinner after a winter of anoxic submergence. However, the amount of calcium lost was less than 4% of the total and not enough to appreciably weaken the shell, nor was the loss great enough to reliably detect by measuring the shell calcium of anoxic turtles.

I decided instead to attempt to prove that the shell was actually the source of the calcium by measuring the buffer properties of shell samples removed from the animals. I performed these experiments with the aid of several undergraduate students: Ryan Armstrong, Zach Goldberger, and Susanna Visuri. Our goal was to simulate what happens in the turtle when its blood is acidic and causes buffer release from the shell. We incubated powdered shell in a solution that was similar to the turtle's own blood plasma but was set at a low pH, similar to the blood of an anoxic turtle. Adding shell powder to the acid solution tended to make the solution pH rise in the alkaline direction, but we held the pH at its preset value by automatic titration with a hydrochloric acid (HCl) solution. The results of this in vitro study provided important insights into how the buffer release mechanism actually works in the animal (Jackson et al., 1999). First, we observed that the rate at which buffer (mainly calcium carbonate) was released (that is, how fast we had to titrate) was directly related to the acidity of the solution. This suggests that the shell releases buffer passively in response to low pH. Second, we were able to measure the rate at which CO_2 was generated in the chamber, and this rate, when compared to the rate of acid titration, matched what would be expected if carbonate was the released buffer anion. Third, we did not detect any phosphate in the solution, so we concluded that the major mineral of bone, calcium phosphate, was not broken down. Finally, the concentrations of calcium and magnesium in the bathing solution increased during incubation to levels similar to those we had observed in the living anoxic turtle. These results provided direct evidence to support our hypothesis that calcium and magnesium carbonate are released from the shell of an anoxic turtle in response to greater acidity of the blood.

To my surprise, however, it turned out that the release of buffer into the blood is not the only way, or even the most important way, that the turtle's shell contributes to managing its acid load.

Lactate Uptake by Shell

On rare occasions in science, one is able to make an unexpected or serendipitous discovery. I now recount one modest instance of this kind in my research. I was conducting a study with the objective of learning how lactate molecules distribute within the turtle's body during anoxia and what the fate of the lactate is during recovery from anoxia. Once again I benefited from the assistance of undergraduate students, in this case Vanessa Toney and Shintaro Okamoto (Jackson et al., 1996). We made painted turtles anoxic at room temperature by submerging them in water for six hours. After two hours of anoxia, we injected a small amount of a specially concocted lactate molecule that had a radioactive isotope of carbon (^{14}C) included in its structure. The presence of the isotope permitted us to detect small amounts of lactate or its metabolic products, such as CO_2, by the use of an instrument (a scintillation counter) that measures emitted radioactivity. The turtles in one of the study groups were anesthetized immediately after the six hours of anoxia, and blood and tissue samples were collected and subsequently tested for ^{14}C activity with the scintillation counter so that we could figure out where the lactate was located. In order to collect the tissue samples, we had to remove the plastron using a hacksaw, and in the process considerable shell powder was generated. On a whim, I decided to test some of the shell powder for ^{14}C activity, although this was not part of the original plan. To our amazement, we discovered quite high ^{14}C activity in these shell samples. Because a turtle while anoxic is unable to metabolize lactate to another compound, we could safely assume that all of the ^{14}C activity we detected in the shell powder was still part of lactate molecules. To verify this assumption, we performed a chemical analysis of the shell powder using a method specific for lactate and got the same result. Lactate had accumulated in the shell, a totally unexpected finding.

The next year I followed up on this finding by studying turtles under anoxic conditions at lower temperatures (Jackson, 1997). At these temperatures much longer anoxic periods are possible because of slower metabolic rates. At a low temperature, turtles can also accumulate higher levels of lactate in their blood. In this experiment I measured lactate in shell samples and in plasma, using only the chemical analytic method. Once again I observed high concentrations of lactate in the shell samples after anoxia at both temperatures. In this study, I also measured lactate concentrations in the liver, heart, and skeletal muscle, which permitted an estimate of the distribution of lactate in the turtle's body at the end of the anoxic period. This calculation yielded very surprising results. At each temperature tested (3°C and 10°C), only 20% of the total body lactate was inside cells, even though this is where all of the lactate is produced. About 36% of the total lactate was in the extracellular fluid (blood plasma and the interstitial fluid, the fluid that bathes the cells), and, remarkably, 44% of the lactate was in the shell. This calculation did not include skeletal bone not connected to the shell which, as subsequent work showed, takes up lactate in the same way the shell does. Had this skeletal bone been included as well, the contribution of bone generally would have been even greater, approaching 50% of the total lactate in the body.

Prior to this observation, no one, to my knowledge, had described this phenomenon of lactate sequestration in bone, and now I discovered that in a painted turtle with high concentrations of circulating lactate, almost half of all of the lactate in its body was in its bone. All of this lactate was produced within the cells of the body, but about 80% of the lactate left the cells, and most of this 80% ended up in bone. A personal disappointment is that I was never able to establish what the physical state of lactate is in the shell, but my hypothesis is that the lactate forms a complex with calcium as I earlier discovered occurs in the blood (Jackson and Heisler, 1982).

The shell and skeleton of the anoxic turtle, therefore, contribute to lactic acid buffering in two ways: first, by releasing buffer into blood to supplement extracellular buffering capacity and, second, by taking up lactic acid, buffering it, and sequestering the lactate in some combined form (Figure 6.4). Using these two distinct mechanisms, the shell and skeleton of the anoxic turtle buffer as much as 75% of the total lactic acid produced. Although other animals also utilize bone buffering, the magnitude of the response is dramatically greater in the turtle. Both mechanisms involve exchange between the blood and the shell and should, therefore, require an adequate flow of blood to the shell. This was shown to be the case by Stecyk et al. (2004), who found that about half of the blood pumped by the heart of an anoxic red-eared slider at 5°C was delivered to the shell.

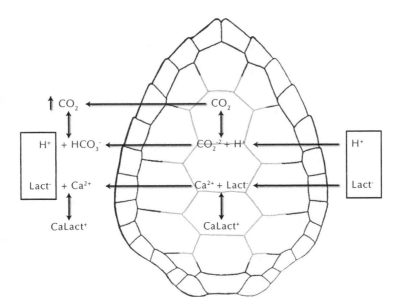

Figure 6.4 Diagram illustrating the two ways that the turtle's shell neutralizes lactic acid. To the left is the release of $CaCO_3$ into the blood to neutralize the acid there. To the right is the movement of lactic acid into the shell, where the neutralization occurs. In both mechanisms, lactate is shown combined with calcium.

In no other animal do we see lactate levels that approach what we have seen in the anoxic turtle, and the turtle is only able to survive such a massive acid load because of the correspondingly massive contribution of buffering from its shell and skeleton. The inescapable conclusion is that the turtle's shell is crucial for its ability to tolerate long periods of anoxia. Without the supplemental buffering supplied by its shell, the turtle's capacity to survive anoxia would probably be similar to the other reptiles studied by Belkin (1963). The turtle's large mass of bone and far greater mineral reserves confer a distinct advantage to this animal in dealing with the consequences of anoxia.

Anoxia Tolerance in Other Turtles

Nearly everything I have written about thus far in this chapter regarding cold, anoxic turtles has derived from observations made on the painted turtle, *Chrysemys picta.* How representative is the painted turtle of turtles in general? Recall that Belkin (1963) found that among reptiles, turtles are exceptional in how long they can live without oxygen, and I can now argue that the possession of a shell is a major reason they are exceptional. All turtles have shells, and all that have been studied have considerable tolerance to anoxia. However, none seems to match the painted turtle. Why? The answer may involve all of the various traits that I have discussed, including the depression of metabolic rate, the abundance of stored fuel in the form of glycogen, and the effectiveness of the shell in buffering lactic acid.

The softshell, *Apalone spinifera,* is a good species to consider first. This pancake-shaped turtle lives in a variety of climates in the United States, from semitropical areas to northern New England. As mentioned in Chapter 5, its leathery shell and perfused skin enhance its aquatic gas exchange, enabling it to survive in a fully aerobic state when submerged at a low temperature in aerated water. However, if the water has no oxygen, then the softshell can barely last two weeks at 3°C. Now it is important to emphasize that surviving two weeks without oxygen is no mean feat, but it does fall well short of the several months endured by the painted turtle. Recall, though, that a key characteristic of the softshell is reduced mineralization of its shell. This turtle simply lacks the enormous buffering reserve of the painted turtle and is unable to cope with high levels of lactic acid (Jackson, Ramsey, Paulson, Crocker, and Ultsch, 2000).

The extent of shell mineralization is not the only factor, however. Many turtles with well-mineralized shells fare worse in anoxia than does the painted turtle. Musk turtles and map turtles, for example, both of which extend into northern regions in the United States, where winters are severe, can remain anoxic at 3°C for "only" one to two months. No simple explanation for their poorer performance compared to the painted turtle is apparent, but experimental evidence points to a more rapid increase in blood lactate during anoxia, suggesting that metabolic depression is not as great, plus less effective buffering by the shell. The buffering deficiencies included less calcium release in relation to lactate increase, and less lactate uptake into the shell (Jackson et al., 2007). The seemingly modest deficiency in anoxia tolerance has substantial ecological significance for these turtles. Unlike the painted turtle, they may not survive if they hibernate in a pond that becomes anoxic for a significant fraction of the winter. On the other hand, these same species have effective aquatic gas exchange, so the selection of overwintering sites that have adequate dissolved oxygen is quite important. Available ecological data suggest that these turtles take these lessons to heart. Recall, for example, that the map turtles in Vermont sit at the bottom of a rapidly flowing river, frozen at the top, but well oxygenated in the cold water below.

Is anoxia tolerance an ancestral trait that evolved back in the Triassic or Jurassic periods when turtles first appeared? Did turtles become trapped at times in stagnant ponds with heavy vegetation at the surface? Or is it a trait that evolved later in species that routinely encounter cold winters with an ice cover that prevents breathing? Gordon Ultsch, Carlos Crocker, and I compared the responses of three species of tropical Southern Hemisphere turtles, *Elseya novaeguineae*, *Emydura subglobosa*, and *Pelomedusa subrufa*, to our familiar painted turtle, *Chrysemys picta*, when all were subjected to six hours of anoxic submergence at 20°C. We found that the turtles of all four species survived this experiment and fully recovered. The painted turtle had less acidosis and its lactate increased less than the others, suggesting a lower metabolic rate, but the significant finding was that all turtles demonstrated excellent anoxic capabilities (Crocker et al., 1999). An important aspect of this result is that the three tropical species are all Pleurodires, or side-necked turtles, a suborder of the turtle lineage that split off from the major branch that led to the Cryptodires, including the painted turtle, some 200 million years ago (Gaffney et al., 2006; see Figure 1.5). This common response to anoxia that we observed in representatives from both groups, including a similar pattern of ionic changes, as described earlier, strongly suggests that anoxia tolerance is a very ancient trait in turtles that can be traced back to the emergence of this reptile.

My strong suspicion is that anoxic survival is a trait that co-evolved with the turtle's shell because of the crucial contribution the shell makes to coping with the acidosis that accompanies anoxia.

Concluding Comments

I was honored in 1999 by being invited to deliver the August Krogh Distinguished Lecture at the annual meeting of the American Physiological Society in Washington, D.C. The subject of my lecture was the same as the subject of this chapter. You may recall from Chapter 1 that August Krogh was a distinguished Danish physiologist and Nobel Laureate from the early twentieth century who, among many notable contributions, enunciated what we call the Krogh Principle, which states, in essence, that for every physiological problem an ideal animal exists somewhere for studying that problem. At the start of my lecture, I quoted a passage from a classic monograph of Krogh's (1941), which stated:

> Crocodiles, turtles, and many tortoises, living in water, dive regularly and many of them are stated to stay for hours or even days under water.... It is evident that the metabolism must be either greatly reduced or become mainly anaerobic. No really quantitative work on the diving of reptiles has been made so far, although some of them should be very suitable for the purpose (79).

Indeed they are, Professor Krogh, and paramount among them is the ordinary painted turtle, an extraordinary animal, and an exquisite exemplar of the Krogh Principle.

7

THE HEART OF A TURTLE

I once did an experiment in which I periodically sampled arterial blood from a turtle submerged in water at room temperature. My purpose was to observe how rapidly the blood oxygen was depleted during a breath-hold period. Because the turtle had breathed shortly before submergence, the oxygen level was high at the beginning of submergence but became progressively lower in the first several samples I took. This was expected, because the turtle's tissues were consuming oxygen and the blood oxygen was not being replenished by breathing. The next sample, however, surprised me. It showed an increase in blood oxygen. How could this happen? The turtle had not breathed. Where did the oxygen come from?

This seemingly incongruous observation can be explained by the curious structure of the turtle's heart, one of the most fascinating aspects of this animal's biology and an object of considerable interest and study by animal physiologists.

Blood Shunting in Turtle Hearts

At the time I made this observation I had a good idea how it could occur, but a subsequent, more thorough study on the same species I studied, the red-eared slider *(Trachemys scripta),* by Warren Burggren and Graham Shelton at the University of East Anglia in the United Kingdom, provided an unequivocal answer. Their published experiment was similar to my unpublished one. They sampled arterial blood at regular intervals and measured oxygen partial pressure of the samples. But they made an important additional measurement. They surgically placed a tube in the lungs, and each time they took a sample of blood they also collected a sample of lung gas and measured oxygen partial pressure in the gas samples as well. In addition, their study was on turtles diving of their own accord, whereas my turtle did not volunteer to dive. Their paper (Burggren and Shelton, 1979) included striking results from a particular turtle that held its breath voluntarily for three hours (Figure 7.1). For the first hour of the breath hold, blood PO_2 fell steadily as expected but, remarkably, lung PO_2 remained nearly unchanged at its initial high, pre-dive level during this same period. The situation changed in the next two sampling periods, as blood PO_2 reversed its course and rose significantly, similar to my observation, whereas lung PO_2 abruptly fell. The same pattern was repeated again: first blood PO_2 fell as lung PO_2 remained steady, and then blood PO_2

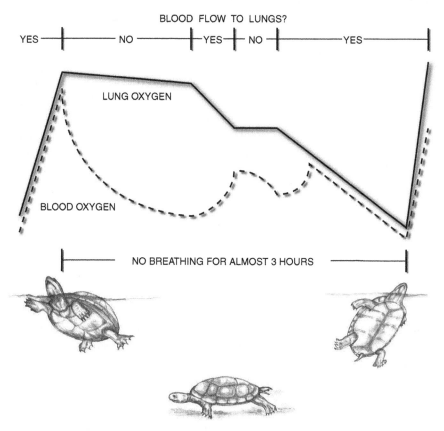

Figure 7.1 Changes in lung and blood oxygen during a three-hour period of breath holding in a red-eared slider turtle (from Burggren and Shelton, 1979; see text for details).

climbed as lung PO_2 decreased. For the final hour or so, lung and blood oxygen decreased steadily in parallel.

Burggren and Shelton's (1979) interpretation of this curious observation, which fit with the prevailing understanding of the circulatory design of turtles, was that during the first hour, when lung O_2 was steady, no blood was being circulated to the lungs, and the O_2 consumed by the tissues was derived solely from the circulating blood. At the end of that first hour, the blood vessels to the lungs opened and allowed blood to flow there; consequently, the O_2-depleted blood returning to the heart from the tissues became oxygenated once again, as O_2 moved from the lung gas into the blood. Blood flow to the lungs was then shut off again for a time before once again reopening.

If a cardiologist discovered that the blood flow patterns in your heart behaved in this fashion, then you would be in need of some serious cardiac surgery. In the human heart, unlike the turtle heart, both atria and ventricles are completely divided by septa into right and left sides. Venous blood flows from our tissues to the right side of the heart, first into the right atrium and then to the right ventricle. From the right ventricle, the blood is pumped to the lungs and back to the left side of the heart through what is called the pulmonary circulation. The blood then passes from the left atrium to the left ventricle, and from there it is pumped out the aorta to the tissues and once again back to the right side of the heart through what is called the systemic circulation. The pathway of blood is a continuous one, passing from the left side of the heart to the systemic circulation, back to the right side of the heart, from there to pulmonary circulation, back again to the left side, and around and around for a lifetime. The separation of the two sides and the presence of one-way valves ensure that blood follows only this path in only this one direction. The volume of blood that the left ventricle pumps into the systemic circulation per minute is commonly called the cardiac output, and from the aforementioned description it is apparent that the right ventricle must pump about the same cardiac output every minute to the pulmonary circulation. If you held your breath, simulating the diving turtle, then your blood and lung PO_2 would each decrease steadily, because flow always continues in the pulmonary circulation. What happened in the turtle heart is called a shunt and is a normal functional characteristic for a turtle but would be considered pathological in a human heart.

In a turtle heart, as in a human heart, the atria are separated from each other by a septum, but in turtles the ventricle is not completely divided;

it has no septum that completely seals off the right side from the left side, as in the human heart. Some fraction of the blood returning to the right side of the heart from the systemic circulation can be pumped right back out into the systemic circulation, bypassing the pulmonary circulation. This is called a right-to-left shunt. An extreme version of this type of shunt was what Burggren and Shelton (1979) observed during the first hour in their diving turtle. Likewise, some of the blood returning to the left side of the heart from the pulmonary circulation may be directed back to the lungs, a pattern called a left-to-right shunt.

Before the last half of the twentieth century, the consensus among biologists was that the heart structure of turtles, and of lizards and snakes whose hearts have a similar structure, was an intermediate stage in the evolution from the heart of fish, which can be considered the ideal design for a water breather, to the heart of a mammal or a bird, which is ideal for an air breather. In both of these "ideal" designs, blood with the highest oxygen concentration leaves the gas exchanger (gills or lungs) and heads out to the tissues without mixing with blood deficient in oxygen, and blood with the lowest oxygen returns from the tissues and flows to the gas exchanger without mixing with oxygen-rich blood. Here I am ignoring those fish with air-breathing organs, which although numerous are vastly outnumbered by fish that lack such an organ. Turtles and their reptilian cousins, according to this earlier view, have made some progress along the path to perfection from fish to mammal/bird, but they are still saddled with structural shortcomings that have not yet been fully corrected. This analysis has some intuitive appeal, but it fails miserably on a couple of counts and is now firmly rejected. First, how long should it take for natural selection to provide the turtle with an improved heart model with a completely divided ventricle? These animals have been around for some 220 million years, seemingly ample time for a new, improved heart design to evolve that would improve fitness. That this has not happened suggests that a turtle may actually be better off with a turtle heart than with a human-type heart. The turtle's heart, imperfect though it may appear to our "mammalocentric" eyes, is, according to the modern view, a suitable heart for this animal and its way of life. Second, the turtle ventricle is not a passive, simple mixing chamber. Instead, it is a highly complex structure that, on the one hand, can function very much like a human heart by sending largely oxygen-rich blood to the tissues and oxygen-poor blood to the lungs and, on the other hand, can vary the proportions of total cardiac flow going to the two circulations over a wide range, even to the extreme of sending no blood whatsoever to the lungs, as observed by Burggren and Shelton (1979). This regulatory flexibility is intriguing and suggests that these shunting patterns serve some useful and important functions in the life of a turtle.

Does Shunting Serve Useful Functions?

Does the ability to shunt blood in the heart improve a turtle's fitness and improve its chances for survival? This seemingly simple question has proven quite difficult to answer unequivocally. James Hicks and Tobias Wang (1996) critically evaluated various suggested functions that were current in the field and argued that none was completely convincing.

The most commonly observed shunting pattern in turtles is a right-to-left shunt (R-L shunt) that occurs during breath holding, as described earlier by Burggren and Shelton (1979), followed by a left-to-right shunt (L-R shunt) when breathing resumes. Recall that a R-L shunt occurs when venous blood returning to the heart with low oxygen and high CO_2 bypasses the lung and flows back out into the systemic circulation, and a L-R shunt is when O_2-rich blood from the lung returns immediately to the lung. Although several possible functions have been proposed, the most commonly suggested reason for a R-L shunt is to restrict blood flow to the lung when the lung is not being ventilated with air. This idea derives from the recognized importance in human lungs for blood flow distribution and air flow distribution in the lungs to be properly matched for optimal gas exchange. An example of mismatching for us would be a blocked airway to part of a lung that prevents that part from receiving any inspired air. Ideally, no blood should be sent to that part of the lung because no gas exchange can occur there. Similarly, a turtle that is underwater and not breathing should not need blood flow to its lungs, but when it returns to the surface and resumes breathing, then it would important for blood to once again flow to the lungs. Just as breathing is intermittent and occurs only when the turtle is at the surface, so too lung blood flow is intermittent and occurs only when the turtle surfaces to breathe. The L-R shunt during breathing, according to this same reasoning, ensures that sending excess blood to the lungs when they are supplied with air enables adequate gas exchange to occur. These ideas sound reasonable, but upon closer consideration and actual testing, they do not measure up.

One problem with the R-L shunt hypothesis during routine diving is that most of the oxygen in a turtle's body when it holds its breath is located in its lungs. If it needs to hold its breath for very long, then the turtle must be able to tap into this oxygen store, and the only way it can do so is by sending blood to the lungs. This is exactly what Burggren and Shelton (1979) observed. The question is why was blood flow to the lungs cut off at all in the turtles they studied? Why not just send blood there all of the time? What difference does it make? Assuming that the rate at which the tissues of a diving turtle consume oxygen is the same whether they shunt blood or not, then it should make no difference whether the flow to the lungs is continuous or intermittent. Yet the possibility exists that bypassing the lungs initially, and letting blood oxygen fall rapidly, has some advantage for the turtle. One suggestion is that the accelerated drop in blood oxygen that results from R-L shunting in a diving turtle may depress metabolism and thereby extend the time available underwater before oxygen runs out. According to this idea, blood is periodically directed to the lungs to keep the blood oxygen in a range where it can adequately supply the tissues while also keeping it low enough to depress metabolism (Hicks and Wang, 1999).

A L-R shunt during breathing has been proposed to be necessary to permit adequate gas exchange during the relatively brief episodes of breathing between dives. However, the rationale for this hypothesis has also been challenged recently in a study by Wang and Hicks (2008). These investigators surgically instrumented red-eared slider turtles in such a way that they could prevent L-R shunting during episodic breathing by restricting flow through the pulmonary artery. Their important finding was that gas exchange was unaffected despite the reduced pulmonary blood flow and a persistent R-L shunt during breathing. They also failed to observe a depression in metabolism during the diving period, which is significant because this was the first test of this idea on fully awake, behaving animals. So the value of intracardiac shunting in turtles during routine diving and intermittent breathing, both the R-L shunt during the dive and the L-R shunt during breathing, has been seriously questioned.

Routine diving by a freshwater turtle at summer temperatures, however, is probably almost always within what is called its aerobic dive limit; that is, the turtle terminates its dive and returns to the surface to breathe before the O_2 runs out. As we have just seen, how helpful shunting is under these conditions is not clear. A very different situation exists, though, during winter months in northern latitudes, when turtles annually experience a far more extended period of breath holding, as discussed in Chapters 5 and 6. Under these conditions, R-L shunting can have a more obvious value. Turtles that spend the winter in bodies of water that freeze may endure weeks or months without breathing, far beyond the time when all of the O_2 in the lungs has been taken up by the blood. When all of the O_2 is gone, blood flow to the lungs is truly no longer necessary for gas exchange. Also, as discussed in Chapter 2, the lungs of a turtle holding its breath for this long will gradually collapse. When collapse is complete, pumping blood to the lungs is not only unnecessary but also difficult, because the resistance to blood flow in a collapsed lung is high and the heart would have to work hard to push the blood through. In other words, maintaining pulmonary blood flow under theses conditions would be energetically costly and of little functional value. Ideally, only enough blood would be sent to the lungs to sustain the living cells there, but this would be a small fraction of the normal cardiac output. If a turtle had a completely divided heart like ours, then it would have no option other than to send all of its cardiac output through the lungs, undoubtedly at considerable energy cost. But with the capacity for intracardiac shunting, the turtle is able to shut off most or all of the flow to the lungs. This would seem to represent a clear value that shunting has for a turtle, at least for a turtle that overwinters in frozen ponds. It could also represent a strong selective pressure to keep the shunting option available.

We cannot easily identify with a turtle holding its breath for many weeks in a frozen pond, and it is hard to believe that we share an experience that is in any way comparable to this. But as developing fetuses in utero, we have all experienced a situation similar in important respects to that of an overwintering, submerged turtle. Within the womb, oxygen delivery into fetal blood occurs not in the lungs but in the placenta, where the O_2-poor fetal blood comes into close diffusion proximity with the O_2-rich maternal blood. In an ice-covered pond, as already discussed in Chapter 5, a turtle obtains whatever O_2 it can, not from its lungs but by diffusion from the surrounding water into the blood circulating through its skin. Both the oxygenated blood flowing from the placenta of the fetus and the blood with added O_2 leaving the skin of the overwintering turtle return to the respective hearts, mixing along the way with O_2-poor blood returning to the heart from the metabolizing tissues. When the blood reaches the hearts of both the human fetus and hibernating turtle, most of this blood is shunted directly back out into the systemic circulation, bypassing the collapsed lungs. In other words in each case a major R-L shunt exists.

The shunt pathways, however, differ in the fetal human heart and in the turtle heart. The fetal heart has a completely divided ventricle but has connections elsewhere that provide short circuit pathways for blood to bypass the lungs. The right and left atria of the fetal heart are connected by an opening called the foramen ovale, and the fetal pulmonary artery and aorta are connected by a vessel called the ductus arteriosus. Some of the returning blood is sent to the lungs, but just enough to supply the metabolic needs of the cells of the lung tissue. Most of the fetal venous blood is shunted over to the left side through the foramen ovale and ductus arteriosus. In the turtle, in contrast, the shunting can all occur within the undivided ventricle, presumably because of the high resistance to flow in the collapsed lungs as well as the constriction of blood vessels in the O_2-depleted lungs. The uninflated lungs of a fetus, like the collapsed lungs of a submerged turtle, are not functioning as a gas exchanger and only need a modest blood flow to supply their metabolic needs. But at the dramatic moment when a newborn baby makes its first appearance into the world, gets the slap on the back, and takes its first breath, the full cardiac output must be directed to the lungs to initiate gas exchange there. Both short-circuit pathways close up soon after birth, and the adult blood flow pattern of complete separation of pulmonary and systemic circulation is established.

A similar event must occur when a turtle takes its first breath after a long winter of submergence, although this has never been studied experimentally. Unlike the well-attended arrival of a newborn human baby, the first spring breath of an overwintering turtle is a rarely observed event. The expected sequence of events, however, is that the emerging turtle reinflates its lungs, restoring its gas exchange function, and pulmonary blood flow is reestablished. In the turtle, no structural feature is altered to make this happen. Instead, flow to the lungs must increase largely because the resistance to flow decreases dramatically due to lung inflation and due to an increase in oxygen pressure within the lungs that stimulates pulmonary blood vessels to dilate. Blood, like water, follows the path of least resistance, so the distribution of blood leaving the heart is influenced by the relative resistances of the two circuits. In a collapsed lung, pulmonary resistance is high and most blood is sent out to the tissues. Once the lung inflates and breathing resumes, pulmonary resistance is low and lung blood flow increases.

I must acknowledge that the description I just gave of the cardiovascular events occurring in a hibernating turtle is largely conjectural, because studies of central cardiovascular shunting in turtles have not been conducted at low winter temperatures. Instead, investigators have made their observations in the comfortable, summerlike environments of their laboratories. The likelihood of my description being accurate, however, is rather high.

How Shunting Occurs

So far I have discussed two of the main objectives of physiologists who study cardiovascular shunting: first, *when* shunting occurs, and second, *why* shunting occurs. Now we turn to the question as to *how* it occurs. In a classic study on red-eared sliders, published in 1966, Fred White and Gordon Ross at UCLA described a R-L shunt during diving and a L-R shunt during breathing and attributed the changes in distribution to resistance changes in the systemic and pulmonary circuits, an important precedent for my previous postulation for the cold turtles. Later, Shelton and Burggren (1976) came to the same conclusion when they observed that pressure changes during a heartbeat were the same everywhere in the ventricle. Their interpretation was that no distributing mechanism existed within the heart itself, but that shunting depended on the relative resistances of the pulmonary and systemic circulations. This mechanism for cardiac shunting, according to this view, is analogous to a parallel electrical circuit in which more current flows through the path with lower resistance. It is important to recognize, though, as of course these authors also did, that the turtle heart is not simply distributing blood to the two circulations on the basis of their relative resistances, but that the heart is also effectively separating the blood so that O_2-rich blood goes primarily to the tissues and O_2-poor blood goes primarily to the lungs. The interior heart structure must in some way enable this separation to occur, and some workers in the field believe that the organization of this structure also contributes to cardiac shunting (Hicks et al., 1996).

The ventricle of a freshwater turtle is considered to have three distinct chambers, but there is continuity between them during at least part of each cardiac cycle. One chamber, called the cavum arteriosum, is comparable to our left ventricle in that it receives O_2-rich blood from the lungs via the left atrium and pumps it out to the tissues via the two aortic arches. A second chamber, called the cavum pulmonale, is comparable to our right ventricle in that it receives O_2-poor blood from the tissues via the right atrium and pumps it out to the lungs via the pulmonary artery. Unlike our heart, however, these chambers are not completely isolated from each other by a septum, so the potential for mixing exists. Mixing is minimized by two structures. First, a muscular ridge isolates the cavum pulmonale during the

121

contraction phase of the heart (systole) so that the O_2-poor blood in this chamber is pumped directly into the pulmonary artery. Second, during the filling phase of the heart (diastole) the cavum arteriosum is isolated by the opening of the valve that allows O_2-poor blood to flow in from the right atrium. The complicating factor in the turtle heart is the third chamber, called the cavum venosum. During filling, O_2-poor blood from the right atrium passes through this third chamber into the cavum pulmonale. During the subsequent contraction, O_2-rich blood from the cavum arteriosum passes through this same third chamber on the way to the aortic arches (Figure 7.2).

These flow patterns have two consequences for shunting. First, residual O_2-poor blood remains in the cavum venosum (the third chamber) after filling and must be flushed out during contraction before the O_2-rich

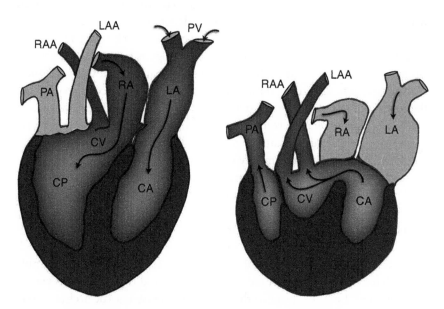

Figure 7.2 Diagram depicting postulated blood flow pattern in the turtle heart. During diastole (left), low O_2 blood returning to the heart from the systemic circulation flows into the right atrium (RA) and thence to the cavum pulmonale (CP) via the cavum venosum (CV). At the same time, O_2-rich blood from the lungs flows through the left atrium (LA) into the cavum arteriosum (CA). During systole (right), the CP pumps blood directly into the pulmonary artery (PA), whereas the CA pumps blood into the two aortic arches, the right aortic arch (RAA) and the left aortic arch LAA) via the CV.

blood from the cavum arteriosum. This constitutes a R-L shunt. Second, some residual O_2-rich blood must remain in the cavum venosum following contraction, and this blood must be flushed into the cavum pulmonale during the next filling phase. When this residual cavum venosum blood is pumped out into the pulmonary artery during the next contraction, it constitutes a L-R shunt. Because of the obligatory flushing of the intervening cavum venosum, some shunting in both directions must always occur, but it may be minimal if the volume of this chamber is small relative to the other two chambers. This suggests that an overall scheme for cardiac shunting in turtles consists of an obligatory shunting pattern due to the functional anatomy of the ventricle and a regulatory mechanism governed by resistance changes in the two circulations that can preferentially send blood to the lungs or to the tissues.

The movement of blood within the heart of a turtle is clearly a complex and perplexing issue. It has not been easy to prove that any of the suggested functions are real or rather just-so stories that we tell ourselves that seem to make sense. One can imagine an experiment that could help us find a way out of this conundrum. Suppose we could surgically "fix" the turtle heart so that it behaved just like a human heart, with a complete separation between the right and left sides of the ventricle? How, if at all, would this compromise the turtle's fitness? This surgical procedure is unfortunately not a realistic prospect because of the complexity of the turtle's heart anatomy. However, a similar surgical intervention is possible in the heart of an alligator.

The alligator and its crocodilian relatives have a unique heart structure. The ventricle is completely divided by a septum, just as in the hearts of birds and mammals. However, the alligator still retains the potential for a R-L shunt because, like other reptiles, it retains two aortic arches leaving the heart, and one of these arches, the left aortic arch, emerges from the right ventricle. If right ventricular pressure is high enough, then oxygen-poor blood can be pumped into this aortic arch, bypassing the lung and producing thereby a R-L shunt. If this arch were tied off where it emerges from the right ventricle, then no R-L shunt would be able to occur. The alligator heart would now function exactly like a mammalian heart, with complete separation of oxygen-rich and oxygen-poor blood. What difference, if any, would it make for the alligator if it were unable to shunt blood?

Investigations are under way to answer this question. Hicks and coworkers have surgically prevented shunting in young alligators and caimans with the object of observing the effects on various aspects of their physiology as they develop. So far, these researchers have not observed any effects of this surgical procedure on growth or exercise capacity after eighteen months (Vorhees et al., 2009) or on dive patterns after twenty-one months (Eme et al., 2009). They have, however, observed higher ventricular blood pressures and a significant enlargement of the ventricle in the animals with the shunt pathway occluded. In another study, Farmer et al. (2008) have recently shown that surgically altered juvenile alligators unable to shunt have reduced maximal rates of stomach acid secretion following a meal compared to sham-operated controls that still retain the shunting capacity. These investigators hypothesize that the shunt pathway, which is normally open after feeding, directs acidic blood with high carbon dioxide to the stomach, where it promotes higher rates of acid secretion.

These surgical interventions are very promising new approaches to understanding shunting and may help resolve some of the issues related to the retention of this capability. Of course, these studies may tell us a great deal about alligators but not necessarily very much about turtles. Conclusive proof of the functional significance of shunting in turtles may not be possible, although it is certainly tempting to assume that some of the suggested functions are vital. Nevertheless, we are still left with the undeniable fact that turtles have been highly successful for tens of millions of years, and we may assume that turtles have had much the same cardiovascular system that we see today throughout their history. If this system is not optimal, then it is certainly adequate, and any selective pressure to change it has not been very strong.

Calcium and the Anoxic Heart

In Chapter 6 I discussed the remarkable ability of freshwater turtles to survive without O_2. For anyone who thinks mainly about human physiology and medicine, much of what was discussed has to be almost unbelievable. One aspect that is worth reemphasizing because it concerns the heart is the manyfold increase in the concentration of blood calcium of the anoxic turtles. Recall that calcium carbonate is released from the turtle's shell and skeleton to help buffer lactic acid that results from anaerobic metabolism. Because the anoxic turtle is essentially a closed system with little or no kidney function, calcium accumulates in the blood to very high concentrations, up to 50 mmol per liter, some twenty-five times its normal level. This is astounding because calcium is perhaps the most tightly controlled element in our body fluids, so even modest increases in the plasma calcium of a patient can be devastating.

I pondered this result and became particularly intrigued by how the turtle's high calcium level might influence its heart function. It is well known that calcium is essential for cardiac contraction in vertebrates. When a human heart cell is excited electrically in the normal way, ion channels, which are highly selective for calcium, open in the cell membrane. Calcium then diffuses through these channels into the heart cell from the fluid surrounding the cell, for two reasons: first, because the concentration of ionized calcium is much lower inside the cells, and solutes tend to diffuse down their concentration gradient, and second, because the negative electrical charge within the cells favors the inward movement of positively charged ions such as calcium. This influx raises the level of intracellular calcium, which has two actions in our heart cells: first, it triggers the release of even more calcium from internal stores in a structure called the sarcoplasmic reticulum (SR), and second, it combines, along with the SR calcium, with a protein called troponin, and this combination enables the heart muscle to contract. As discovered in the nineteenth century by English scientist Sidney Ringer, calcium must be present in the fluid outside of the heart cell or the heart muscle will not contract. The sequence of events just described for the human heart is the same in turtle heart cells, except the SR calcium release is of minor importance in turtles (Galli et al., 2006).

Some calcium outside of the cell is therefore essential for a turtle's heart to contract, but what about an excess of calcium? In humans, high blood calcium (hypercalcemia) is decidedly bad and can produce cardiac arrhythmias or even stop the heart, but the effect must be different in an anoxic turtle. I already knew from our whole animal studies that the turtle's heart continues to beat while anoxic, despite high levels of circulating calcium, albeit at a slow rate and with reduced pressure generation compared to a heart with oxygen. But what effect might the calcium have?

I decided to violate the familiar scientist's axiom that "a week in lab can save you an hour in the library" by consulting the literature, and I found that a number of studies of mammalian hearts had shown that acidosis can interfere with the entry of calcium into heart cells during excitation and can also compete with calcium at sites within the heart cell, such as the binding site on troponin (e.g., Williamson et al., 1976). On this basis I hypothesized that elevated blood calcium in the anoxic turtle, although primarily associated with helping to buffer the lactic acid produced during anoxia, might also serve another important function; it may help minimize the inhibition of heart function caused by the acidosis.

I was able to interest an undergraduate student, Hal Yee, in testing this hypothesis and for his honor's thesis at Brown he conducted a study that showed that elevated calcium did indeed improve the contractile function of isolated atrial muscle exposed to an acid solution (Yee and Jackson, 1984). Subsequent studies in my lab (Wasser et al., 1990) and elsewhere (Overgaard et al., 2005) confirmed this finding in ventricular muscle. Although the improvement in heart function in response to elevated calcium was always modest, it was still a positive effect, and the high calcium certainly did not harm the heart (Jackson, 1987). Yee's study also launched my lab into further investigations of heart function under anoxic and acidotic conditions and eventually to the application of the technique of nuclear magnetic resonance (NMR) spectroscopy.

It was during a sabbatical leave I took in Strasbourg, France, during 1988 that a postdoctoral fellow in my lab, Jeremy Wasser, interested organic chemistry professor and NMR expert Ronald Lawler at Brown to collaborate on a study of isolated turtle hearts. When I returned home we began these studies, and they continued over the next eight or nine years, bringing in additional students, including Karen Inman, Elizabeth Arendt, Patricia Hamm, Cheryl Watson, and Hongyu Shi, as well as a Skidmore College colleague, Roy Meyers, who was a visiting scientist in my lab at the time.

The experimental preparation was somewhat involved and worth describing. After exposing the heart of a turtle, we surgically inserted a tube into its right atrium for filling the heart and inserted a second tube into one of the aortic arches for conducting pumped fluid out of the heart. All other vessels connected to the heart were tied off, and the heart was removed from the animal. In doing so we were careful to preserve the pacemaker region of the right atrium so that the heart would continue to beat spontaneously. We then mounted the heart on a special holder and placed it into a glass NMR tube filled with a salt solution that simulated the turtle's normal body fluids, a type of fluid that to this day is called Ringer's solution, in honor of the scientist, mentioned earlier, who first concocted such a solution. One of the fluid's key constituents was, of course, calcium. We placed the tube with the heart on ice and carried the ice bucket and assorted other equipment across campus to the chemistry building, where the NMR instrument was located. The instrument resembles a giant, fat thermos jug, and it was situated across from a console, where the operating computer was located. The glass tube containing the heart was inserted from the top, down into the depths of the machine, with the long connecting tubes trailing over the top and back down to reservoir bottles and collecting tubes mounted on the outside. Once all connections were made and the experiment was under way, fluid flowed by gravity through one of the long plastic tubes from the reservoir up and then down into the machine, where it filled the heart during the relaxation phase (diastole) of the heart cycle. Another long plastic tube carried the fluid pumped by the heart during the contraction phase (systole) of the heart cycle up and over the top of the machine and down, until it emptied into a graduated cylinder for measurement of cardiac output. The fluid in the supply reservoir could be set at a normal or an acid pH or at high or zero oxygen. After everything was prepared and under way, the normally bare metal wall of the NMR machine was covered with tubes, bottles, tape, and syringes, initially an astounding sight for passersby in the Chemistry Department.

The heart was carefully positioned within the NMR spectrometer in what is affectionately called the "sweet spot," where the excitation pulse of the instrument was focused. A uniform powerful magnetic field is centered on this spot. Our experiments looked at the atomic nuclei of phosphorus (^{31}P), a component of a number of important intracellular compounds such as ATP (adenosine triphosphate). When excited by a pulse of radio frequency radiation of the appropriate frequency, the orientation of the spinning nucleus of the element is changed, producing a transient signal at or near the exciting frequency. The secret of NMR spectroscopy is that the same element produces a signal with a frequency that differs slightly, depending on its particular chemical environment. For example, each of the three phosphorus atoms of ATP has a signal at a slightly different frequency, because each has a distinct chemical environment within the ATP molecule. The intensity versus time signal from the excited nuclei in the object of study is transformed to give a frequency spectrum on which each detectable phosphorus atom is represented by a peak. The peak of a reference molecule (in our case, creatine phosphate) is used as the zero position on the spectrum, and the other atoms (or compounds) can be identified by the distance their peaks are from the reference peak, a separation called the chemical shift. In our experiments we could identify a number of important phosphorus-containing compounds, including ATP, creatine phosphate, and inorganic phosphate. The size of the peaks provided information on the relative concentrations of these compounds. The magnitude of the chemical shift between inorganic phosphate and creatine phosphate, with suitable calibration, provided a measure of intracellular pH. Because measurement of intracellular pH is an important application when studying the phosphorus spectrum of a biological specimen, the NMR machine, when used for this purpose, has been called "the world's most expensive pH meter." One limitation of NMR spectroscopy is that it only works with elements (such as ^{1}H and ^{31}P) that have the required nuclear properties and are in a chemical environment that gives relatively "sharp" peaks that are suitable for analysis. Another problem with regard to phosphorus is that the technique is not particularly sensitive and for ^{31}P requires that compounds be at a relatively high concentration, a limitation that unfortunately excludes many phosphorus-containing compounds in the cell. For interested readers, Wasser et al. (1995) provide a more detailed description of the principles of NMR spectroscopy as applied to biological preparations.

A representative spectrum from a turtle heart is shown in Figure 7.3. Easily identified peaks are inorganic phosphate (Pi), creatine phosphate (CP), and the peaks for the three phosphates of ATP (α, β, and γ). Interestingly, the largest peak in the turtle heart NMR spectrum, the size of which remained unchanged during our experiments on a given heart, is a composite of several compounds called phosphodiesters (PDE). Whereas the functions of ATP, CP, and Pi are well known, we still do not know why these PDE compounds are in such high concentration in the turtle's heart.

Our studies confirmed what had been known to generations of students who used to study turtle hearts in their biology labs: the turtle heart is a durable organ that can continue to beat and pump blood for long periods outside of the body if treated with reasonable care. A more specific finding was that the turtle heart, in this case the heart of the painted turtle, could continue to pump at close to normal strength and cardiac output, even when deprived of oxygen for many hours. This rather extraordinary behavior helps explain how the intact turtle can survive many weeks of anoxia during winter hibernation.

But we also found that a turtle's heart is not invulnerable to the full range of stresses associated with oxygen lack. When the heart was supplied with a solution that had no oxygen (anoxic) and at a low pH (acidotic), a combination that simulates what happens during natural hibernation in a stagnant pond, heart function was severely depressed. From experiments on intact turtles undergoing anoxic submergence, we knew

128

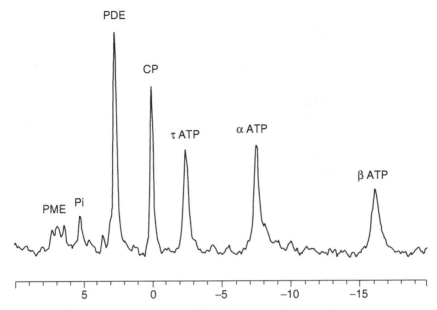

Figure 7.3 An NMR spectrum from a perfused turtle heart. Each peak represents a different phosphorus-containing molecule: PME, phosphomonoesters; Pi, inorganic phosphate; PDE, phosphodiesters; CP, creatine phosphate; and the three phosphates of adenosine triphosphate (ATP), γ, α, and β.

that both heart rate and blood pressure are depressed. Even in this most stressful of conditions, though, when the heart had no oxygen and its pH was very low, the heart continued to beat, although more weakly, and ATP in the cells remained high, only modestly reduced from the normal level. Furthermore, after several hours in this state, when the solution supplying the heart was replaced with one with oxygen and at a normal pH, heart function rapidly returned to normal, demonstrating the resilience of this impressive organ (Wasser et al., 1990).

We also performed experiments in which we stopped all flow to the heart for ninety minutes or more at 20°C, producing a condition known as ischemia, and we found that when flow was restored, the heart recovered rapidly, with no evidence of injury (Wasser et al., 1992). Ischemia in a human heart can permanently damage the region not receiving blood flow and produce what is called an infarct. The anoxia-tolerant turtle heart showed no evidence of such damage.

129

Other extracellular factors, in addition to calcium, that can influence heart function also change in the blood of the whole animal during anoxia. In our original study at 3°C, plasma potassium concentration increased as much as fivefold during anoxia, and in a later study Wasser and I found that the plasma hormones epinephrine (=adrenaline) and particularly norepinephrine (=noradrenaline) increased dramatically during combined anoxia and acidosis (Wasser and Jackson, 1991). Elevated potassium would be expected to depress function because of the effect on the electrical activity of heart cells. The hormones epinephrine and norepinephrine, on the other hand, would be expected to improve function, just as when the heart is activated by these agents during exercise or stress. Sure enough, Nielsen and Gesser (2001), studying strips of cardiac muscle from turtles, confirmed the stimulating effect of epinephrine and the depressant effect of high potassium, but they observed as well that when administered together, the effects canceled each other out.

What is it about the turtle heart that sets it apart from our heart or the hearts of other mammals that cannot survive very long without oxygen? One fundamental difference is that a mammalian heart cannot support its high metabolic requirements using only anaerobic metabolism. Without O_2, the production of ATP cannot keep pace with the use of ATP, and this leads to a fall in the concentration of this essential source of energy for the cell. As a result the heart cannot sustain energy-requiring functions such as contraction and the maintenance of membrane ionic gradients, and this in turn leads to a cascade of events, starting with a massive influx of calcium, the activation of protein-degrading enzymes, and eventually the death of the cells. The turtle heart, in contrast, has an intrinsically lower energy demand than the mammal heart, and under the metabolically depressed anoxic state, the demand is well below its maximum anaerobic energy-producing capacity (Farrell et al., 1994). This means that adequate ATP can be generated to match its rate of ATP consumption, and the heart can continue to function for long periods with no O_2.

Supplying Oxygen to the Heart

As just described, we prepared the turtle heart for our NMR experiments by tying an inflow tube into the right atrium and an outflow tube into an aortic arch, and we supplied the circulating fluid to the heart through these tubes. It is worth noting that this method would not work with a mammalian heart. In the standard preparation to study the isolated heart of a mammal, called the Langendorff preparation, a tube is inserted into the aorta and is aimed toward the heart. This tube supplies oxygen-rich fluid under high pressure in the reverse direction of the usual flow of blood. The fluid cannot enter the left ventricle of the heart because the aortic valve is a one-way valve that only permits outflow from the heart, but fluid can flow into the coronary artery that arises from the root of the aorta. By doing so, it produces fluid flow through the coronary circulation, which is absolutely necessary to supply oxygen and nutrients to the mammalian heart and to keep it functioning outside of the body. The importance of this circulation to our normal well-being is familiar to us all. Blockage of a major artery leading into this network by atherosclerotic plaque deprives the portion of the heart muscle supplied by that vessel of blood and oxygen, and a heart attack will ensue. If the blockage is serious enough, the individual will die; in fact, heart attack is the leading cause of death in the United States. If the condition is not fatal, then flow can be restored by coronary artery bypass surgery, or by opening the occluded vessel using angioplasty or inserting a stent. The important point is that our heart and the hearts of other mammals, birds, and even some lower vertebrates depend critically on an effective coronary circulation. But this is not the universal rule among vertebrates, and it is not the case in freshwater turtles that have been studied in this regard.

A turtle heart, as well as the hearts of many other lower vertebrates, has only a modest coronary circulation that is restricted to the outer compact surface of the heart (Brady and Dubkin, 1964). Most of a turtle's heart muscle depends instead upon diffusion of oxygen and nutrients directly from the blood within the chambers of the heart to supply its needs. The inner portion of the heart is like a sponge through which the blood percolates and makes intimate contact with the muscle cells. The good news for these animals is that they need not be concerned about coronary artery disease, but the potentially bad news is that some of their heart is being supplied with venous blood coming from the tissues, and this blood may not provide an adequate oxygen supply, particularly during exercise.

The circulatory design of a fish exemplifies this problem most clearly, and this is the animal that potentially has the most serious problem. Blood in a fish travels through a single circuit from the tissues through the heart to the gills and back out to the tissues. This design, as noted earlier, is in one sense ideal because the blood leaving the gills with the highest oxygen content in the system travels directly to the tissues that need that oxygen. But unless the fish has a coronary blood supply derived from a post-gill artery, the heart, because it is upstream from the gills, will be supplied with oxygen-poor venous blood flowing from the tissues. This is the case for both chambers of the simple fish heart, the single atrium and the single ventricle. Furthermore, when the fish becomes active and the heart must work harder to pump more blood to the active tissues, the blood that returns from the tissues to supply the heart has even less oxygen because the active muscles remove a greater fraction of the oxygen from the flowing blood. So at the very time when the heart needs more oxygen, the concentration of oxygen in the blood returning to the heart is reduced.

Some active fish, such as tuna, avoid the problem by having a coronary circulation that supplies the heart with oxygen-rich blood. But how do fish that lack coronaries deal with this problem? Colleen Farmer, who was a graduate student in my lab at the time, suggested that fish that are facing this challenge have two options (Farmer, 1997). One is to be sluggish and have little capacity for aerobic activity. The second is to provide a supplemental source of oxygen during activity by using an air-breathing organ to infuse oxygen-rich blood into the blood returning to the heart. This is the option on which she focused.

Air breathing is surprisingly common in fish evolution and, according to Jeffrey Graham (1997), at least 347 species from 49 different families of fish exist today that exploit the air to obtain at least some of their oxygen. A variety of exchange surfaces are used, including the mouth, the stomach, the swim bladder, and true lungs. The common view among physiologists has been that air-breathing organs in fish are principally adaptations for living in bodies of water in which the oxygen can become depleted. Rising to the surface and gulping oxygen-rich air helps the fish survive when aquatic oxygen is low. Colleen argued that air breathing in fish might serve another function as well. She proposed that air breathing is recruited when a fish becomes active, and that supplementing the oxygen uptake in this way increases the level of oxygen in the blood returning from the tissues to the heart. Air-breathing organs in fish are generally situated in the systemic circulation, and the blood draining these organs empties into the general venous blood flowing back to the heart. When such a fish is active and its heart is working harder to pump more blood, air breathing provides another source of oxygen to keep the heart functioning.

Colleen performed an experiment in my lab that provided support for her hypothesis. She built a flow tank (a flume) in which a fish had to swim against the current to maintain its position. Within the flume, fish were free to breathe water with their gills or to come to the surface and breathe air. The species she studied were the garfish, *Lepisosteus oculatus,* and the bowfin, *Amia calva,* both air-breathing fish that have functional gills. Her results were striking. When the fish were at rest, they rarely breathed air and obtained nearly all of their oxygen directly from the water. However, while swimming, they increased their air breathing, and each fish, on average, extracted more than half of its oxygen from the air (Farmer and Jackson, 1998). This experiment does not prove that the fish breathe air exclusively to supplement the heart's oxygen supply during activity, nor does it rule out the importance of low water oxygen as a breathing stimulant, but it is perfectly consistent with Colleen's novel hypothesis.

Compared to fish, tetrapods (amphibians, reptiles, birds, and mammals) have more complex circulatory designs, with separate circulations to the lungs (pulmonary circulation) and to the tissues (systemic circulation). But just as in fish, venous blood returns to the heart from the tissues depleted of oxygen, and O_2 depletion is more severe when the animal exercises. Venous blood low in O_2 fills the chambers on the right side of the heart, whereas the left side of the heart receives blood from the lungs with high oxygen content. How do these animals deal with the potential problem of low oxygen in the right side of the heart? As noted earlier, mammals, birds, and also some reptiles such as crocodilians have a coronary circulation that supplies arterial, oxygen-rich blood to the whole heart, both the left and right sides. Amphibians have effective skin gas exchange that adds additional oxygen to the blood flowing back to the heart from the systemic circulation, similar to the contribution of an air-breathing organ of a fish. But what about turtles that have only slight coronary circulation and that lack effective skin gas exchange at temperatures where they are active? How does a turtle prevent possible right-side heart problems when it is active and venous blood is low in oxygen? Colleen proposed that the left-to-right shunt, which recirculates O_2-rich blood back to the lungs, provides the right side of the heart with the O_2 it needs. This idea is supported by the observation that left-right shunting increases in turtles when they are active (Krosniunas and Hicks, 2003). By shunting oxygen-rich blood to the right side of the heart and back out to the lungs during activity, the right side of the heart is provided with supplemental oxygen to help support its increased pumping.

Concluding Remarks

The heart of a turtle is an impressive organ indeed. Its versatility and resilience certainly exceed that of the human heart. Heart rates in painted turtles or red-eared sliders can range from one beat every ten minutes in a cold, anoxic turtle (Herbert and Jackson, 1985a) to one hundred beats per minute in a warm, exercising turtle (Gatten, 1974). This thousandfold difference far exceeds our usual range of threefold to fourfold. Turtle hearts can continue to beat and pump blood with no O_2 and with the severe disturbances to normal homeostasis that accompany long periods in this state. The human heart fails rapidly in the absence of O_2. The structure of the turtle's heart provides the animal with myriad options for directing its outflow of blood. The turtle heart can direct O_2-poor blood largely to the lungs and O_2-rich blood to the tissues as in the human heart, or it can, in a regulated fashion, shunt blood in either direction, even to the extreme of totally bypassing the lungs and sending the entire outflow to the tissues. What at first may appear to be deficiencies in the heart design of a turtle are more accurately viewed as adaptive features that have been selected for and retained over the millions of years of this animal's evolution. The heart of the turtle, like the animal itself, is an ancient and enduring example of successful design.

8

LIFE IN THE SLOW LANE

The tortoise won the race against the hare not because of its swiftness but because of its persistence. The much faster hare raced ahead but then stopped to rest and fell asleep. The plodding tortoise passed it by and reached the finish line first. As Aesop moralized, "Slow but steady wins the race."

Metabolic Rates of the Tortoise and the Hare

Slow moving and plodding suggest a low rate of metabolism and, sure enough, if you compare a tortoise with a hare, the metabolism of the hare is much higher, whether the two are at rest or racing against each other. The hare is a mammal, a warm-blooded animal that maintains a high body temperature and requires considerable energy to do so. The tortoise is a reptile, a cold-blooded animal with a body temperature similar to the surrounding temperature, and it consequently needs less metabolic energy. Published data reveal that a resting hare consumes oxygen about twenty-five to thirty times faster than a resting tortoise of similar body weight if both are at 20°C. A difference of twenty-five to thirty times is huge. It is approximately the same as the difference between Lance Armstrong's maximal rate of O_2 consumption (VO_{2max}) and his basal metabolic rate. The hare is equivalent to Lance on a breakaway pedaling up a slope in the French Alps, and the tortoise is Lance resting quietly during the off-season.

The greater rate of oxygen usage by the hare must be reflected in all of the processes that deliver oxygen to its metabolizing cells. Oxygen moves from the environment to an animal's cells through a series of steps that resemble a bucket brigade in which a bucket of water passes from person to person until it reaches the fire. If ten gallons of water are dumped on the fire each minute, this requires each member of the brigade to move ten gallons each minute. The first step in the transfer of O_2 is movement of air into the lungs by breathing. Next, oxygen diffuses from the lung gas across a thin barrier into the blood passing through the lungs. The heart then pumps the O_2-containing blood from the lungs to the metabolizing tissues, where again oxygen diffuses the short distance from the tiny capillaries into the cells. Once in the cells, oxidative metabolism occurs within mitochondria, producing ATP, the energy currency of the body. In a resting hare, each step of this pathway must move twenty-five to thirty times as much O_2 as in a resting tortoise, and similar differences exist if both animals are exercising. The hare must breathe much more; it must have a larger and thinner surface in its lungs to facilitate diffusion; it must have a stronger heart that can pump more blood; its blood must have more hemoglobin to carry O_2; it must have a denser capillary bed to provide adequate surface for diffusion to cells; and its cells must possess more mitochondria to utilize the O_2. Furthermore, to provide the fuel for its metabolism, a hare has to ingest twenty-five to thirty times as many calories as the tortoise, which means it must spend much more time foraging for food. Its gut structure and function must be scaled up to process and absorb all of this additional food. In short, the hare operates in the fast lane of life, while the tortoise plods along in the slow lane.

Over the years I have frequently measured the metabolic rates of turtles of various species, and these results were always consistent with the expected low values. However, I also was involved in a study in which the metabolic rate of a hare was determined. During my postdoctoral work with Professor Knut Schmidt-Nielsen, I had the opportunity to join him and others, including my PhD advisor Ted Hammel, in a field study of jackrabbits in southern Arizona (Schmidt-Nielsen et al., 1965). Jackrabbits are bona fide hares of the genus *Lepus*. We carried out this study at a U.S. Forest Service facility at the base of the Santa Rita Mountains in a beautiful desert setting. Jackrabbits were plentiful, but we had to figure out a way to capture them. Because they are active after dark, we collected them at night. Our technique, which was very successful, was to shine a spotlight on a jackrabbit from the back of a pickup truck, which usually froze the animal in its tracks. Two of us would then quietly climb off the truck, circle behind, and capture the rabbit with a hand net. No method is perfect, however, and the downside of this strategy was that those of us doing the capturing had to move rather blindly through surrounding vegetation, which on one occasion resulted in an unexpected encounter with a rattlesnake and, on more occasions than I wish to remember, intimate contact with a cactus called jumping cholla. Months after this expedition, I was still removing cactus spines from my skin. These are among the hazards of field research.

Our study was mainly concerned with the thermoregulatory aspects of this desert-dwelling hare, and we learned much about how this medium-size, non-burrowing animal with quite large ears manages to survive in a hot, dry desert. However, included in our study were measurements of the resting metabolic rate of the black-tailed jackrabbit *(Lepus californicus)*, as estimated by oxygen consumption. Minimum resting O_2 consumption, observed at 20°–25°C for a 2.3 kg jackrabbit, was about 0.6 mL O_2 (g hr)$^{-1}$. The estimated standard O_2 consumption of a turtle or tortoise of the same body weight and also at 20°C, based on an analysis by Bennett and Dawson (1976) of published measurements, is about 0.022 ml O_2 (g hr)$^{-1}$. These values are consistent with the twenty-fivefold to thirtyfold difference mentioned earlier. It is quite remarkable that both animals function well and maintain physiological homeostasis, but the turtle does so at a small fraction of the cost. A sizable fraction of the excess heat produced by the hare is simply to keep its body temperature nearly constant over a wide range of environmental temperatures. Leakier cell membranes in the hare and other mammals allow gradients for ions such as sodium and potassium to dissipate more rapidly and therefore require more energy for pumping ions uphill to sustain these gradients (Hulbert and Else, 1981). In addition, research indicates that mitochondrial membranes in mammals are leakier for protons (H^+), and that energy must be expended to resist this leakage (Brand et al., 1991). The proton gradient in the mitochondria is essential for the generation of ATP for the cells.

The higher rates of energy turnover in warm-blooded animals serve functions other than simply generating more heat. Birds' and mammals' more powerful metabolic engine permits them to achieve much higher levels of activity, and some investigators propose that during the evolution of warm-bloodedness, the regulation of a high body temperature was secondary to a boost in exercise capacity (Bennett and Ruben, 1979). Aesop's fable has staying power, both because of its moral lesson but also because it contradicts both anecdotal and scientific evidence that the hare should have easily outdistanced the tortoise. The tortoise was definitely the underdog in that legendary race.

I actually had a race with a hare during our Arizona study, but this humble event has never been immortalized by a writer of fables. Prof.

Schmidt-Nielsen believed that a human runner could chase down a jack-rabbit because the rabbit, unable to lose body heat fast enough, would become exhausted and stop because of high body temperature. This would happen sooner in the rabbit than in the larger human. To test this idea, Schmidt-Nielsen gathered our whole team during the middle of a hot Arizona day, and we formed a large, loose circle around a likely habitat. Gradually we closed the circle and, sure enough, a jackrabbit suddenly burst forth from within the circle and, as luck would have it, raced past quite near where I was positioned. I took off after it. I had done some running in school and was still in reasonably good physical condition so I was confident that I could provide a good test of Schmidt-Nielsen's hypothesis. My jackrabbit prey behaved somewhat like Aesop's hare; it raced away from me and then stopped. However, this hare did not fall asleep but merely waited until I got close and then took off again. This sequence of stopping and starting was repeated several times over perhaps a half mile, until the rascal darted around a mesquite tree. When I reached the tree, the jackrabbit had vanished. That ended my chase, leaving unresolved the issue we were testing.

In retrospect if a tortoise had been stationed at that critical place in our circle instead of me, then perhaps the outcome would have been different.

Temperature and Metabolic Rate

The comparison of the metabolic rates of the tortoise and the hare, as described earlier, was based on measurements at a standard ambient temperature of 20°C. Choosing this temperature is not unreasonable because it is a temperature commonly experienced by each animal. If the purpose, though, is to compare the intrinsic metabolic rates of each animal, then a 20°C environment biases the outcome in favor of the hare. At an ambient temperature of 20°C, the body temperature of the tortoise would also be about 20°C, but the body temperature of the hare would be close to 37°C, the normal body temperature of a mammal. A fairer comparison would be to study the two animals at the same body temperature because body temperature has a major impact on metabolic rate. To do this, the turtle would have to be in an environment with a temperature of 37°C and allowed to equilibrate there. This could be an abnormally high temperature for some turtles, but for many terrestrial tortoises, such as the gopher tortoise *(Gopherus polyphemus)* of the southern United States or the desert tortoise *(Gopherus agassizii)* of the southwest United States, it would not be a problem. The effect of increasing the turtle's body temperature from 20°C to 37°C would be to raise its metabolic rate approximately threefold to sixfold, reducing the metabolic edge of the hare from twentyfold to thirtyfold to about fivefold to tenfold. The metabolic rates of cold-blooded animals such as tortoises normally speed up approximately two to three times for each 10°C increase in body temperature. Physiologists call this acceleration in rate the Q_{10}. Normal Q_{10} values are in the range of 2 to 3, which means that the rate of metabolism doubles or triples for each 10°C increase in body temperature. Most turtles do not spend much of their lives at temperatures near 37°C, however, so functionally the comparison between the mammal and the turtle is probably more appropriate at 20°C. Incidentally, when we measured the metabolic rate of a jackrabbit at an air temperature of about 37°C, the rate was almost the same as at 20°C because the animal's body temperature had not changed appreciably.

Operating at a high body temperature can be important for a turtle, but because it is a reptile (that is, a cold-blooded animal) its low metabolic rate and poor insulation do not allow it to elevate its body temperature by using its own internal heat production. Instead, a turtle must seek out a warmer environment and allow that environment to heat it up. For this reason, turtles and other so-called cold-blooded animals are more properly called ectotherms, meaning their body heat derives principally from outside their bodies, from the surrounding environment. Hares, like other mammals and birds, on the other hand, are called endotherms because their body heat is mainly derived from within the body by their high metabolic rate. A higher metabolic rate, as well as superior insulation in the form of fur, feathers, or fat, enables an endotherm to maintain a large temperature difference between its body and its environment. The familiar way that an aquatic turtle elevates its temperature is by crawling out onto a rock or other object and basking in the sun. In the summer, this can raise its body temperature to perhaps 30°–35°C and facilitate various bodily processes, such as digestion. Also, with warmer muscles, the turtle can swim more vigorously when it returns to the water. Of course since the water is probably cooler, the turtle will lose this advantage as it equilibrates to the lower water temperature.

The Q_{10} effect works in the other direction as well. If body temperature falls, then so does metabolic rate. When Christine Herbert measured O_2 consumption of painted turtles in my lab at 20°C, 15°C, 10°C, and

3°C, she observed a steady fall in metabolic rate that accelerated as temperature got lower (Herbert and Jackson, 1985b). Between 20°C and 15°C, she observed a conventional decrease in metabolism that was calculated to have a Q_{10} of 2.9; however, between 15°C and 10°C Q_{10} was 4.9, and between 10°C and 3°C the Q_{10} had risen to 8.5. As a result of the high Q_{10} values, the metabolic rate at 3°C was only 6% of the rate at 20°C and one-third the value it would have reached had the rate fallen throughout this temperature range with a Q_{10} of 2.9. At a low temperature, a turtle is not feeding and is very likely hibernating. As discussed in Chapter 6, low metabolism is advantageous under these conditions because it slows the rate at which stored fuel is consumed. Furthermore, if an aquatic turtle is submerged in water, then lower metabolism can more readily be satisfied with O_2 diffusing directly from the water. High Q_{10} at the low end of its temperature range thus has distinct advantages for a turtle, but the trade-off is that the turtle is even more sluggish than usual and more vulnerable to predation, and even more seriously handicapped in any competition with the hare or with the hare's aquatic fellow mammals.

Body Size and Metabolic Rate

It is conventional to express metabolic rate on the basis of body weight (or mass), but the use of body weight as an index for calculating metabolic rate is complicated in a turtle because of the large mass of its shell, a structure that has a comparatively low metabolism. Bennett and Dawson (1976) considered this issue in their review and suggested that perhaps the metabolic rate of a turtle should be based on its body weight excluding the shell. However, when they compared the metabolic rates of turtles to other reptiles, normalized to body weight without making any correction for the shell, they found no significant difference. The turtle's shell does have living cells and does consume oxygen, but it is also likely that the scatter in the data was such that the effect of bone mass differences was obscured.

Turtles are a diverse group of animals and vary greatly in body size. During my career I have worked on individuals ranging from a 4 g painted turtle (Reese et al., 2004) to a 142 kg adult green turtle (Prange and Jackson, 1976). That represents a size range over thirty-five thousandfold, but I did not study the largest turtles. Male Galapagos tortoises can exceed 300 kg, whereas the largest turtles of all, leatherback sea turtles, can reach 700 kg, and a stranded individual weighing over 900 kg has been reported, so the total range for all turtles is even greater, well over a hundred thousandfold. The range of size within a single species of turtle can also be quite large. A green turtle hatchling, at about 30 g, is bigger than a painted turtle hatchling, but the adult we studied at Tortuguero had increased in size from its birth weight by almost five thousand times. And, as discussed in Chapter 1, a leatherback turtle may increase in size by twenty thousand times during its lifetime. By comparison, our own body weight typically increases from a birth weight of perhaps 3–4 kg (7–9 lbs) to an adult weight of about 60–80 kg (130–180 lbs), an increase of only fifteen to thirty times.

When Henry Prange and I made our first trip to Tortuguero, one of our research objectives was to estimate the size dependence of metabolic rate in green turtles. We already knew that in mammals the rate of O_2 consumption per gram of body weight is higher in a small animal than in a large animal. Moreover, Henry had already collected metabolic rate data during both rest and exercise from both green turtle hatchlings weighing about 31 g (Prange and Ackerman, 1974) and from immature turtles weighing about 800 g (Prange, 1976). At Tortuguero we had access to nesting female green turtles weighing well over 100 kg, and this would enable us to define the size dependence from animals of the same species covering a large range of body sizes. At the time we performed the study, data on large turtles were scarce and limited to a single study by Hughes et al. (1971) on the Aldabra giant tortoise that involved individuals spanning a three hundredfold size difference. Our results revealed a significant size effect on metabolic rate. The O_2 consumption of a 142 kg green turtle (Prange and Jackson, 1976) at Tortuguero was eleven hundred times as great as the hatchling; however, the adult weighed forty-five hundred times as much. In other words, each gram of the large turtle consumed O_2 at less than 25% of the rate per gram of the hatchling. Combining these data with the values from the intermediate-size immature green turtles (Prange, 1976), we calculated that the resting metabolic rate increased as a function of body mass to the 0.83 power. Expressed mathematically ($\propto$ is the symbol for "proportional to"):

$$O_2 \text{ Consumption per Gram} \propto \text{Body Mass}^{0.83}$$

Subsequent measurements we made on adult green turtles (Jackson and Prange, 1979) and on immature green turtles (Kraus and Jackson, 1980) agreed closely with the earlier values, and a recent review that summarizes all relevant data for resting green turtles describes a relationship

141

between size and metabolic rate that is very similar to our original value (Wallace and Jones, 2008). To help interpret this mathematical relationship, consider that if the metabolic rate of the green turtles increased in direct proportion to body weight, in other words if a turtle one hundred times as large has a metabolic rate one hundred times as high, then the rate of metabolism would be proportional to body mass to the 1.0 power. The lower power function of 0.83 indicates a progressive decrease in the metabolism of each gram of tissue as the turtles become larger.

German physiologist Max Rubner's pioneering nineteenth-century work on the effect of body size on metabolic rate introduced the so-called "surface law." Based on measurements of dogs of different body sizes, he concluded that metabolic rates varied as a function of body surface area, and that consequently metabolic rate increased as a function of body mass to the 0.67 power. This was a quite reasonable idea because dogs all regulate their body temperatures at about 37°C, and heat transfer between their bodies and their environments occurs across their body surfaces. As an animal gets larger, its surface area increases as the square of the linear dimension, but its mass increases as the cube of its linear dimension, so a surface-dependent phenomenon should increase with the two-thirds power of body mass. However, the subsequent analysis by Kleiber (1947) of a large set of mammals covering a much larger range of body size challenged Rubner's surface law and indicated instead a body mass exponent of 0.75, intermediate between surface dependence (0.67) and a simple linear relationship (1.00). Furthermore, the 0.75 value also is the approximate value for animals that do not keep their temperatures constant. Various ectotherms, including unicellular organisms, as well as endothermic birds, follow the same general pattern (Hemmingsen, 1960). Uncertainties still abound, though, because of deviations of particular groups, like our green turtles, from the standard exponent value. However, even more fundamentally, it is still not clear what physical principles account for the power law describing the change in metabolic rate with body size. Readers who are interested in pursuing the debate surrounding this topic can consult the *Journal of Experimental Biology* 208, no. 9 (2005), where some of the major players in the matter present their arguments and theories.

Exercise and Metabolic Rate

Turtles are more notable for their inactivity than for their activity. Nonetheless, they engage in all of the various pursuits of life, such as foraging, feeding, reproductive behavior, exploration, and escape, all of which demand some increase in metabolic rate above the standard resting rate. Exercising turtles can use both aerobic and anaerobic metabolism. Aerobic actions are more sustainable and fatigue resistant, whereas anaerobic metabolism supports intense effort that can rapidly result in exhaustion and termination of activity. (Think of taking a long walk versus racing to catch the bus.)

As discussed in Chapter 4, Henry Prange and I, in our study of green sea turtles at Tortuguero, recorded increases in O_2 consumption up to ten times the standard resting rate when turtles were hauling about on the beach or engaged in natural nesting behavior. Both of these endeavors are sustainable over a period of hours and are mainly aerobic, although we did detect lactic acid in blood samples from these turtles, which indicates some contribution from anaerobic metabolism. A recent study of nesting green turtles at a beach on the Gulf of Oman has confirmed this finding and reported even higher levels of blood lactate than we observed (AlKindi et al., 2008). Sea turtles commonly engage in long-term activity such as nesting and migration that must be principally aerobic. Short-term bursts of energy that turtles employ to escape from danger have a higher anaerobic component. Robert Gatten (1974) experimentally induced maximal metabolism at 30°C for two minutes in red-eared sliders *(Trachemys scripta elegans)* and measured both O_2 consumption and total body lactate. The lactate value enabled Gatten to calculate anaerobic metabolism expressed as mmoles of ATP produced per gram. The oxygen consumption rate was also converted to equivalent ATP production. During the two-minute period of intense activity, the turtles produced 62% of their ATP using anaerobic metabolism.

Major reliance on anaerobic metabolism during burst activity is characteristic of reptiles generally (Bennett and Ruben, 1979), and its importance is greater at shorter durations of intense activity. In a recent study by Hancock and Gleeson (2008), desert iguanas *(Dipsosaurus dorsalis)* ran at maximal intensity for fifteen seconds and then the various sources of ATP production supporting this effort were determined. Remarkably, O_2 consumption (aerobic metabolism) accounted for only 2.5% of the total. Anaerobic glycolysis, the pathway leading to lactate, was the major contributor of ATP and was responsible for 65% of the total. The authors attributed the remaining 29% to the compound creatine phosphate. Creatine phosphate concentrations are typically high in skeletal muscle, and this molecule is the immediate source of high-energy phosphate to replenish ATP as

143

it is broken down to ADP (adenosine diphosphate) during burst activity. Creatine phosphate is a major source of energy for human sprinters during a 100 m dash or swimmers during a 50 m swim when the athletes may not breathe at all throughout the event. When a freshwater turtle swims vigorously to an underwater refuge, an activity that probably only takes a few seconds, we can safely assume that O_2 consumption plays a relatively minor role and that anaerobic processes predominate.

Based on Gatten's (1974) study, however, and on our own perception of their behavior, turtles have less explosive anaerobic capacity than some other reptiles and are more in line with the capacity of mammals. It is worth returning to the comparison between ectotherms and endotherms and reconsidering the relative magnitudes of metabolic activity. As emphasized, an endotherm such as a hare has a much higher resting metabolism than a tortoise and can also achieve much higher maximal levels of O_2 consumption. This translates into greater sustained running speed by the hare, the reason the hare was the favorite to win the race.

However, what if maximal, unsustainable running speeds are compared, speeds at which the full anaerobic capacities of the animals are recruited, and instead of a tortoise (not a great runner) we consider instead a lizard as our ectotherm? A mammal's top sprinting speed is only about twice its maximal sustainable (aerobic) speed. For example, the Olympic 100 m champion sprints at a speed of 37.5 km hour^{-1}, less than twice the speed of the victorious marathoner who runs his 42 km race at 20 km hour^{-1}. In contrast, lizards have maximal speeds that are ten to thirty times their sustainable aerobic speeds (Bennett and Ruben, 1979). This is no surprise to anyone who has tried to chase down a lizard on a warm day. If the lizard can reach safety in a few seconds, then one is not likely to catch it. The difference is the huge anaerobic burst capacity of the lizard.

These considerations help answer what otherwise might seem to be puzzling questions: How can a cold-blooded animal catch, kill, and eat a warm-blooded prey? Should not the endotherm's superior metabolic capacity enable it to outfight the ectotherm? The answer to the second question would be yes if the ectotherm (and for our example let us now choose a crocodile) could only use its limited aerobic metabolism. However, when a crocodile attacks, it does so with a burst of anaerobic effort and with powerful, fast contracting muscles that most mammals are no match for. Mammals (even humans) should be able to outrun a crocodile, assuming they see it coming, but if it surprises them, seizes them in its powerful jaws, and drags them into the water, then they do not have much of a chance. When my wife and I visited Australia, the locals up in the Northern Territory regaled us with stories of man-eating crocodiles and tried to convince us that their favorite food was American tourists. Although we were suspicious of this latter bit of intelligence, we were nonetheless quite wary when walking along the shores of billabongs.

Feeding and Metabolic Rate

Crocodiles tend to eat rather large meals (not always American tourists, thankfully) at infrequent intervals, and associated with their large meals is a sizable increase in metabolic rate. The increase, called "specific dynamic action," is due to the intake, digestion, absorption, and assimilation of the meal and is well known in human physiology. Because we tend to eat several times a day, the metabolic effect in humans and other mammals is rather modest, an increase of only 25%–50% over our standard metabolic rate. The increase can be much larger, however, for many cold-blooded animals that have longer fasts and larger meals. Consider the American alligator, *Alligator mississippiensis* (Busk et al., 2000). When young alligators, weighing 9 kg or less, consumed rats or mice equivalent to about 7.5% of their body weight, their metabolic rate (O_2 consumption) rose within a few hours, reached a peak that was approximately four times higher than before feeding, and remained significantly elevated for at least four days. An even more dramatic metabolic response occurs in the sidewinder rattlesnake, *Crotalus cerastes,* a sit-and-wait predator that consumes large, infrequent meals. In a study by Secor et al. (1994), snakes were fasted for at least three weeks and then fed meals that averaged about 26% of their body weight. Once again, it is instructive to relate that to our own experience. If you weigh 160 pounds, you would have to ingest in a short time a pile of food weighing over 40 pounds. Even the gluttons who win the hotdog-eating contests would be humbled by that challenge. But snakes do this for a living, and some are known to consume more than their body weight in a single meal.

Within a day after the meal provided by Secor and colleagues, O_2 consumption of the sidewinders rose significantly and reached a level eight times the fasting rate by the end of two days. This approaches the increase in metabolism that these animals can achieve during maximal aerobic exercise. Associated with the processing of this large meal was a dramatic

145

building of intestinal structure and function. Following completion of the digestive process, which lasted about two weeks, the intestine atrophied, and the authors concluded that the gut would remain in this reduced state until the next meal. This same pattern of high metabolic rate and gut proliferation occurs in other ectotherms that eat large, infrequent meals (Wang et al., 2006). The cycle of growth and atrophy is not restricted to the gut. In the burmese python, heart size increased by 40% within two days of consuming a meal amounting to 25% of the snake's body mass (Andersen et al., 2005). The heart then returned to its resting size when digestion was completed. This remarkable observation reveals a rate of heart growth unprecedented in human physiology and, as the authors state, ". . . could provide an attractive model for investigating the fundamental mechanisms that lead to cardiac remodelling and ventricular growth" (38).

Most turtles, even predatory ones such as snapping turtles, feed more frequently, thus their response to feeding should be less dramatic than that of inveterate binge eaters such as snakes and crocodiles. Several studies have shown this to be the case, although the turtles do exhibit some of the same responses when fasting is prolonged. For example, Secor and Diamond (1999) fasted snapping turtles, musk turtles, and red-eared sliders for one month and then fed them meat equivalent to 5%–11% of their body weight. They then observed that, similar to the response of the other reptiles, O_2 consumption increased significantly, up to three and a half times the pre-feeding rate. They also calculated that over the course of the digestive process, the cumulative increase in metabolic rate consumed about 20% of the total energy contained in the meal. Unlike that which occurs in rattlesnakes, however, the digestive tract of the turtles did not increase either morphologically or functionally in response to the meal. These turtles apparently keep their digestive systems in a state of readiness, even during a period of fasting, which fits with their normal strategy of frequent feeding. Secor and Diamond (1999) did not rule out the possibility that other turtles may differ in this regard, or that the species they studied may decrease the size and function of their guts during even longer periods of fasting, such as during hibernation.

Metabolic Scope

As is clear from what has already been said in this book about metabolic rate, turtles can exhibit quite different rates of metabolism depending on their circumstances (see Figure 8.1). The data in this figure are derived

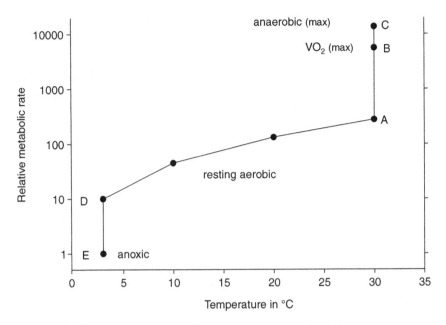

Figure 8.1 The metabolic rate of freshwater turtles as affected by temperature, activity, and anoxic submergence. Note that the overall change from the highest to the lowest metabolic rate is over thirteen thousandfold. (See text for details.)

from two studies, one in the high temperature range on red-eared sliders, by Gatten (1974), and the other in the low temperature range on western painted turtles, by Herbert and Jackson (1985a). It is convenient to begin with the turtle at rest at 30°C (point A in Figure 8.1). Engaging in some active pursuit at that temperature or feeding can increase the rate of O_2 consumption severalfold from the standard rate, and the study by Gatten (1974) showed that maximal effort can raise the rate of aerobic metabolism approximately twentyfold (point B in Figure 8.1) and the total rate by about fiftyfold (point C in Figure 8.1). Thus at 30°C, aerobic production of ATP is responsible for only about 40% of the total during maximal effort, the balance being attributable to anaerobic metabolism. A fiftyfold range of metabolic rate is impressive, but many mammals and birds can match this.

Turtles, though, can greatly extend this metabolic range by reducing their metabolism. Even at a high temperature, near 30°C, certain species of freshwater turtles will experience a summer drought that will necessitate a period of inactivity or estivation. Kennett and Christian (1994) reported marked reductions in O_2 consumption in the Australian long-neck turtle, *Chelodina rugosa,* to well below the predicted standard (or basal) metabolic rate when they studied this animal during simulated estivation in the laboratory. However, in a field study on a closely related species, *Chelodina longicollis,* Roe et al. (2008) did not observe a decrease. Other studies have also failed to measure a decrease below the standard resting metabolic rate during estivation (e.g., Ligon and Peterson, 2002), but certainly the quiescent state of these turtles keeps their rates of metabolism well below the routine rates of behaving animals and thus serves to conserve energy during a time of stress.

Because turtles are ectotherms, their body temperature can fall well below 30°C either daily or seasonally. If a painted turtle sitting on a log basking on a summer day with a body temperature of 30°C slips into the water and slowly equilibrates with water at 20°C, then the turtle's resting rate of O_2 consumption will fall twofold or threefold, as shown in Figure 8.1. At 20°C, the turtle once again has a scope of metabolic rate associated with activity and feeding, although the maximal and minimal rates shift downward from rates at 30°C. When winter arrives and the turtle cools even more, to perhaps 3°C, the turtle's resting metabolic rate falls still farther (point D in Figure 8.1). Recall that this decrease is accentuated by a large increase in the Q_{10} at a low temperature. The rate at 3°C is close to the lowest rate of aerobic metabolism that a turtle can achieve and is over five hundred times less than the VO_2max at 30°C. This approximates the overall aerobic metabolic scope of a freshwater turtle, its maximal range of oxygen consumption, and vastly exceeds the tenfold to twentyfold scope of a human at constant temperature.

If the water's surface is still unfrozen, then the turtle may swim to the surface to breathe and will have a limited scope for increasing its metabolic rate due to exercise. Feeding does not occur at this temperature, so no metabolic effect due to the specific dynamic action of ingested food will contribute. If the surface freezes and the turtle can no longer breathe air, then metabolism may fall farther as the turtle sinks into its hibernation state, with only direct uptake of O_2 from the water to support its limited range of aerobic metabolism. In addition, if depletion of O_2 in the water requires the turtle to rely solely on anaerobic metabolism to produce ATP then its metabolic rate will sink still farther to only about 10% of its fully aerobic rate at 3°C (point E in Figure 8.1). The turtle has now reached what is perhaps its lowest viable rate of ATP production. It is now decidedly situated in the "slow lane," metabolically speaking.

The profoundly reduced metabolic rate of the anoxic turtle at 3°C is more than four orders of magnitude (over thirteen thousand times) lower than the peak level that a turtle can achieve during maximal effort at 30°C. This is an enormous metabolic rate range for a single animal but is probably routinely experienced each year of their lives by painted turtles that live in temperate climates. In summer, a painted turtle basking in the sun is an alert, active animal that can achieve a high rate of ATP production as it moves swiftly to a safe refuge if disturbed. The same turtle during the following winter rests quietly at the bottom of its frozen pond, barely moving, its metabolic flame turned down to a barely detectable setting. As the seasons pass, this animal shifts smoothly between these metabolic extremes, matching its metabolic state to the environmental condition and following a pattern that has persisted for millions of years. Safely housed within its shell with its adaptable physiology and behavior, the turtle is a survivor and an amazing evolutionary success story.

EPILOGUE

When I decided to stop doing research and close down my lab, I still had four painted turtles in my animal facility. They were healthy animals that had never been used in an experiment. I had to decide how to dispose of these four beautiful turtles. The obvious option, which is generally the rule for surplus experimental animals, was to euthanize them humanely by administering an anesthetic overdose. But this was not the way I wanted it all to end. I procrastinated and finally took the dilemma home and consulted with my wife Diana. I explained the situation to her. I told her I had to do something with the turtles but said my options were limited. I could euthanize them, but I absolutely did not want to do that. I could release them in a local pond, but they were western painted turtles, not native to Rhode Island, and this would be ecologically incorrect. So what to do? Diana asked me where the turtles came from, and I told her Wisconsin.

Diana looked me straight in the eye and calmly said, "Then let's drive to Wisconsin and release them there."

Momentary pause.

"Okay."

This conversation took place toward the end of June 2006. Less than three weeks earlier we had been out west and had climbed the Grand Canyon on our fortieth wedding anniversary. Were we ready for another trip in that direction? Of course we were. It was summer, with a full week ahead of us with nothing on our calendars, and we were in an adventurous mood. So from the moment I said "okay" it was totally decided, and we never reconsidered. We simply started preparing for the trip.

Painted turtles ordinarily do not travel very far. The home range of painted turtles living in a marsh in Michigan was about 1–2 ha (2–5 ac) (Rowe, 2003). Ranges can be greater if the turtles live in a river system, if they must migrate because their pond dries up, or if a female wanders far afield to find a suitable nest site. In more restricted habitats, such as studied by Rowe (2003), a turtle, during its entire life, may travel only 200 m (660 ft) or less in various directions from the center of its range. But my four turtles had already traveled way beyond this modest limit, much of it in an airplane that carried them 1,000 miles from Wisconsin to Rhode Island. According to the fossil record, painted turtles as a species have existed in North America for perhaps 5 million years (Starkey et al., 2003); the airplane has existed for a mere hundred years, so for 99.8% of their evolutionary history, traveling by air would not have been a possibility for painted turtles.

Now these turtles were about to take another long trip, this time strictly by car. This journey began on Sunday morning, June 25, when we transported them by car, from the lab to our home, in a plastic container. We added water and placed the container on our porch. Several neighborhood children came to admire our visitors. Later that day we drove to our son Tobey's home in Arlington, Massachusetts, where we created an enclosure for the turtles with rocks gathered from their garden. Our grandchildren, Chloe and Seth, admired the turtles, albeit from a safe distance. We gathered inside for dinner, but I periodically went outside to check on the turtles. After dinner, when I went out to check on the turtles again, I found to my dismay that three had managed to breach the walls of their enclosure. I called for assistance, and we soon recovered two of the escapees who were nearby. The third turtle was nowhere to be seen, but after an agonizingly long search (actually only a few minutes), Tobey discovered the errant turtle in the grass, heading toward the front of the house. This was a reasonable direction for an aquatic turtle to head because across from the house flows the Mystic River, but it was a potentially tragic direction because a rather busy thoroughfare lies between the house and the river. To reach the river, the turtle would have had to cross that road. Like the airplane, the automobile is a recent phenomenon in the long history of this species, and like other turtles, painted turtles have not learned how to avoid this mysterious threat to their well-being. But thanks to Tobey's sharp eyes, the water-bound turtle was captured before it reached the road, and once again all four turtles were safely under our care. We returned home to Providence, Rhode Island, and kept the turtles overnight in their plastic tank.

The next morning we loaded our personal effects and the empty plastic tank into the trunk of our black VW Beetle convertible and placed the turtles in the backseat in a covered cardboard box with wet newspapers. We set off down the road at about 8 a.m. toward I-95. Although I am not able to enter deeply into the mind of a turtle, I am fairly certain that our four traveling companions were not too happy about their situation. I wished that I could somehow help them understand the purpose of our trip, but they are creatures of the present moment, and the present moment was not terribly good for them. They were enclosed in a dark, damp container and subjected to the movements of the car and the roar of its motor. The convertible top, however, was up, and it remained up for the whole trip, so ambient noise was muted somewhat. A friend of ours, Lucy Stevens, made a charming but fanciful drawing of us tooling down the road with our chelonian cargo (Figure E.1).

We drove down I-95 to New Haven, Connecticut, where we cut over via Danbury to I-84, the highway that took us out of the state, across a section of New York State, and into Pennsylvania. In Scranton we turned south on I-81 and then west on I-80, the highway we would travel all the way to Chicago. We stayed the first night at a Quality Inn in western Pennsylvania, near the Ohio border. As far as we could tell, there were no restrictions about keeping reptiles in our room. The turtles may have appreciated the somewhat calmer environment of the motel and also the opportunity to leave the cardboard box and swim around in the plastic container. But they still had a long way to go and another full day of riding in the backseat of the Beetle.

We set off again the next morning, still on I-80, on the final leg of our mission that would take us across Ohio and Indiana, past Chicago, and then up into Wisconsin, where western painted turtles live. It was a long drive and not always a pleasant one because of the abundance of tractor trailers plying that route, particularly in the vicinity of Chicago. At times the large trucks hemmed us in on all sides. We hunkered down in the shell of our VW Beetle surrounded by these behemoths, not unlike a small turtle making its way in a hostile world. But any doubts about what we were doing out on this inhospitable highway were dispelled by the occasional rustling in the backseat as the turtles moved about in their box, reminding us of our goal and keeping us focused and in high spirits.

When we reached Wisconsin we were at the eastern limit of where the western painted turtle, *Chrysemys picta bellii* (the subspecies of our passengers), resides. Four subspecies of painted turtles have been described that are distributed widely throughout North America (Figure E.2); indeed, the species *Chrysemys picta* has the largest geographical distribution of any North American turtle (Starkey et al., 2003). In Rhode Island,

Figure E.1 A fanciful depiction of the trip west to release the turtles (drawing by Lucy Stevens).

where our trip began, the eastern painted turtle *(Chrysemys picta picta)* is the resident subspecies. Its range lies along the eastern seaboard, from Maine down into Georgia. At about the time we entered Pennsylvania on our journey west, we left the territory of the eastern painted turtle and entered into the range of the midland painted turtle, *Chrysemys picta marginata*. This subspecies ranges across Pennsylvania and most of New York State, up into Ontario and southern Quebec, down through western Virginia, into eastern Tennessee, over into Kentucky, and up into eastern

153

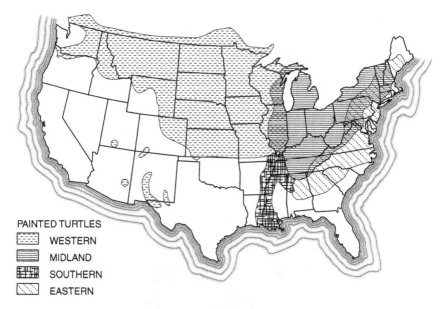

PAINTED TURTLES
- WESTERN
- MIDLAND
- SOUTHERN
- EASTERN

Figure E.2 Distribution of the four subspecies of painted turtles within North America (adapted from Conant and Collins, 1991).

Illinois and southeastern Wisconsin. The range of the western painted turtle extends from Wisconsin across the northern United States and southern Canada, all the way to the West Coast and south into Wyoming, Colorado, and New Mexico. The fourth subspecies, the southern painted turtle, *Chrysemys picta dorsalis,* lives in the southern part of the Mississippi Valley, well outside of the itinerary of our road trip. The different subspecies can be distinguished in the field by the characteristic patterns of their shells (Carr, 1952). Because they are all the same species, these different subspecies can mate and produce intergraded offspring at sites where their ranges overlap.

We entered Wisconsin on I-94 late Tuesday afternoon and stopped at a Welcome Center on the highway to get information. We wanted to drive into the countryside and find a suitable pond to release our turtles, and we hoped the center could help us. When we asked about the location of bodies of water where turtles might be found, however, it was clear that the friendly folks behind the counter had never been confronted with this question. Had we asked about golf courses, amusement parks, accommodations, and the like, they would have had brochures and useful advice, but for an ideal spot to release some turtles, they had no clue. After some discussion, they agreed that if we headed west we would definitely find many ponds and lakes, so we left and followed their advice. Sure enough, after some driving, we started passing some rather large lakes, but we needed to know whether there were resident turtles. We obviously needed to talk to some knowledgeable locals.

After more driving we saw a sign for a bait and tackle shop, which seemed like a perfect location for finding someone who knew the local ponds and their inhabitants. Unfortunately, the shop was closed but a roadside restaurant and bar was close by and it was "happy hour." We parked the Beetle and I walked into the bar. The room was dimly lit and as I entered the four or five men who were sitting with their backs to me at the bar turned and looked at me and their conversation stopped. I was clearly a stranger.

"Does anyone know a local pond that has a population of painted turtles?" I asked.

The looks I got at that point suggested that they thought I had just dropped down from another planet. I briefly explained my mission and they were more confused than ever, but they invited me to have a seat and tell them more. Diana came in and joined us. As the conversation went on, despite their amused comments about the wackiness of our purpose, it became evident that we had hit pay dirt. These were all local people who knew the country and the creatures that live there. Amazingly, one of them had even been down to the Pacific coast of Costa Rica, where he had volunteered in a turtle tagging effort. I told him about my experiences at Tortuguero on the Caribbean coast. Bob, the most talkative of the lot, knew a fair bit about turtles and their habits and told of finding groups of hibernating snapping turtles in a nearby pond. Our new friends treated us to a beer and regaled us with stories of their turtle experiences, among other topics. I was careful in our conversation not to mention that I am a New England Patriots fan, since we were in Green Bay Packer country (the establishment was even called the "Packer Inn").

Eventually, Bob said he knew just the place for our turtles and gave us directions. We thanked him and the others and, assuming that we understood his directions, we drove off to find the pond. However, we were soon lost and had obviously not gotten the directions straight after all. So we found our way back to the bar and sheepishly admitted our failure to find our way. That probably confirmed their already low regard for these strange easterners. But Bob, being a very helpful gentleman, said he would lead us to the pond on his motorcycle. So we got back in the car and followed him on the circuitous path to a small lake where, it turned out, his cousin lives with his family. They came out to greet us, and we all walked around to the back of the house, Diana carrying the box with the turtles, where we found a beautiful body of water. An arm of the small lake came close to their backyard and was covered with duckweed, although the surface of the main body of the lake was clear.

The moment of truth for our road trip had arrived. Diana set the cardboard box, with the turtles rustling around in the damp newspaper, on the grass beside the pond. She opened the box and we looked at the turtles that were still as uncertain as ever about what on earth was going on and about what their fate might be. I then had the exhilarating experience of picking them out of the box one by one and placing them on the edge of the pond. Each turtle hesitated briefly, getting its bearings, and then moved, cautiously at first but then more purposefully, into the water and soon disappeared beneath the surface. Diana released the last of the four, and now all were safely in the pond, out of sight. The box was empty. The turtles were back in their normal habitat.

Turtles are patient and persistent but also inscrutable. What is going on inside the "mind" of a turtle? When I suggest that our turtles were concerned or uncertain or whatever, I am imposing on them my own sense of what they may feel. In fact I have no idea what they experience. Nonetheless, I cannot help but believe that these four turtles, when they entered that pond, experienced some measure of relief and pleasure. At the very least they may have appreciated no longer being frustrated by never finding a way out of their box. I know for sure that I felt happy to see them return to their natural setting, and I would like to believe that they felt the same way. I sincerely hope they still do.

REFERENCES

ACKNOWLEDGMENTS

INDEX

Ackerman, R. A., and F. N. White. 1979. "Cyclic carbon dioxide exchange in the turtle *Pseudemys scripta*." *Physiol. Zool.* 52: 378–389.

Aesop. 1984. *The Aesop for children with pictures by Milo Winter*. Chicago: Rand McNally

AlKindi, A. Y. A., A. A. Al-Habsi, and I. Y. Mahmoud. 2008. "Changes in plasma levels of adrenaline, noradrenaline, glucose, lactate and CO_2 in the green turtle, *Chelonia mydas*, during peak period of nesting." *Gen. Comp. Endocrinol.* 155: 581–588.

Andersen, J. B., B. C. Rourke, V. J. Caiozzo, A. F. Bennett, and J. W. Hicks. 2005. "Postprandial cardiac hypertrophy in pythons." *Nature* 434: 37–38.

Belkin, D. A. 1963. "Anoxia: Tolerance in reptiles." *Science* 139: 492–493.

Bennett, A. F., and W. R. Dawson. 1976. "Metabolism." In *Biology of the reptilia, vol. 5, Physiol. A,* ed. C. Gans and W. R. Dawson, 127–223. New York: Academic Press.

Bennett, A. F., and J. A. Ruben. 1979. "Endothermy and activity in vertebrates." *Science* 206: 649–654.

Berkson, H. 1967. "Physiological adjustments to deep diving in the Pacific green turtle (*Chelonia mydas agassizii*)." *Comp. Biochem. Physiol.* 21: 507–524.

Bickler, P. E., and L. T. Buck. 1998. "Adaptations of vertebrate neurons to hypoxia and anoxia: Maintaining critical Ca^{2+} concentrations." *J. Exp. Biol.* 201: 1141–1152.

Bowen, B. W., A. B. Meylan, and J. C. Avise. 1989. "An odyssey of the green sea turtle: Ascension Island revisited." *Proc. Natl. Acad. Sci. USA* 86: 573–576.

Brady, A. J., and C. Dubkin. 1964. "Coronary circulation in the turtle ventricle." *Comp. Biochem. Physiol.* 43: 119–128.

Brand, M. D., P. Couture, P. L. Else, K. W. Withers, and A. J. Hulbert. 1991. "Evolution of energy metabolism: Proton permeability of the inner membrane of liver mitochondria is greater in a mammal than in a reptile." *Biochem. J.* 275: 81–86.

Broderick, A. C., M. S. Coyne, W. J. Fuller, F. Glen, and B. J. Godley. 2007. "Fidelity and over-wintering of sea turtles." *Proc. Roy. Soc. B* 274: 1533–1538.

Broderick, A. C., B. J. Godley, and G. C. Hays. 2001. "Trophic status drives interannual variability in nesting numbers of marine turtles." *Proc. Roy. Soc. B* 268: 1481–1487.

Broderick, A. C., B. J. Godley, S. Reece, and J. R. Downie. 2000. "Incubation periods and sex ratios of green turtles: Highly female biased hatchling production in the eastern Mediterranean." *Marine Ecol Prog. Ser.* 202: 273–281.

Buck, L. T., P. W. Hochachka, A. Schön, and E. Gnaiger. 1993. "Microcalorimetric measurement of reversible metabolic suppression induced by anoxia in isolated hepatocytes." *Am. J. Physiol.* 265: R1014–R1019.

Burggren, W. W., and G. Shelton. 1979. "Gas exchange and transport during intermittent breathing in chelonian reptiles." *J. Exp. Biol.* 82: 75–92.

Burke, A. C. 1989. "Development of the turtle carapace: Implications for the evolution of a novel bauplan." *J. Morphol.* 199: 363–378.

Busk, M., J. Overgaard, J. W. Hicks, A. F. Bennett, and T. Wang. 2000. "Effects of feeding on arterial blood gases in the American alligator, *Alligator mississippiensis.*" *J. Exp. Biol.* 203: 3117–3124.

Butler, P. J., W. K. Milsom, and A. J. Woakes. 1984. "Respiratory, cardiovascular and metabolic adjustments during steady state swimming in the green turtle." *Chelonia mydas. J. Comp. Physiol. B.* 154: 167–174.

Cameron, J. N. 1989. "Acid-base homeostasis: Past and present perspectives." *Physiol. Zool.* 62: 845–865.

Carr, A. 1952. *Handbook of turtles: The turtles of the United States, Canada, and Baja California.* Ithaca, NY: Comstock.

———. 1967. *So excellent a fishe: A natural history of sea turtles.* Garden City, NY: Natural History Press.

———. 1975. "The Ascension Island green turtle colony." *Copeia.* 1975: 547–555.

Carr, A., and P. J. Coleman. 1974. "Seafloor spreading theory and the odyssey of the green turtle." *Nature* 240: 128–131.

Carr, A., L. Ogren, and C. J. McVea. 1980. "Apparent hibernation by the Atlantic loggerhead turtle off Cape Canaveral, Florida." *Biol. Conserv.* 19: 7–14.

Carrier, D. R. 1987. "Lung ventilation during walking and running in four species of lizards." *Exp. Biol.* 47: 33–42.

Cebra-Thomas, J., T. Fraser, S. Sistla, E. Estes, G. Bender, C. Kim, P. Riccio, and S. F. Gilbert. 2005. "How the turtle forms its shell: A paracrine hypothesis of carapace formation." *J. Exp. Zool.* 304B: 558–569.

Conant, R., and J. T. Collins. 1991. *A field guide to reptiles and amphibians of Eastern and Central North America.* New York: Houghton Mifflin.

Crews, D., J. M. Bergeron, and J. A. McLachlan. 1995. "The role of estrogen in turtle sex determination and the effect of PCBs." *Environ. Health Perspect.* 103 (Suppl. 7): 73–77.

Crocker, C. E., R. Feldman, G. R. Ultsch, and D. C. Jackson. 2000. "Overwintering behavior and physiology of eastern painted turtles (*Chrysemys picta picta*) in Rhode Island." *Can. J. Zool.* 78: 936–942.

Crocker, C. E., T. E. Graham, G. R. Ultsch, and D. C. Jackson. 2000. "Physiology of common map turtles *(Graptemys geographica)* hibernating in the Lamoille River, Vermont." *J. Exp. Zool.* 286: 143–148.

Crocker, C. E., G. R. Ultsch, and D. C. Jackson. 1999. "The physiology of diving in a north temperate and three tropical turtle species." *J. Comp. Physiol. B* 169: 249–255.

Daniels, C. B., and S. Orgeig. 2003. "Pulmonary surfactant: The key to the evolution of air breathing." *News Physiol. Sci.* 18: 151–157.

Dessauer, H. C. 1970. "Blood chemistry of reptiles: Physiological and evolutionary aspects." In *Biology of the reptilia, morphology C, vol. 3,* ed. C. Gans and T. S. Parson, 1–72. New York: Academic Press.

Donohoe, P. H., T. G. West, and R. G. Boutilier. 1998. "Respiratory, metabolic, and acid-base correlates of aerobic metabolic rate reduction in overwintering frogs." *Am. J. Physiol.* 274: R704–R710.

Douse, M. A., and G. S. Mitchell. 1990. "Episodic respiratory related discharge in turtle cranial motoneurons: In vivo and in vitro studies." *Brain Res.* 536: 297–300.

Dunson, W. A. 1967. "Sodium fluxes in fresh-water turtles." *J. Exp. Zool.* 165: 171–182.

Eme, J., J. Gwalthney, J. M. Blank, T. Owerkowicz, G. Barron, and J. W. Hicks. 2009. "Surgical removal of right-to-left cardiac shunt in the American alligator *(Alligator mississippiensis)* causes ventricular enlargement but does not alter apnoea or metabolism during diving." *J. Exp. Biol.* 212: 3553–63.

Ernst, C. H., and R. W. Barbour. 1989. *Turtles of the world.* Washington, DC: Smithsonian Institution Press.

Farmer, C. 1997. "Did lungs and the intracardiac shunt evolve to oxygenate the heart in vertebrates?" *Paleobiol.* 23: 358–372.

Farmer, C. G., and D. C. Jackson. 1998. "Air-breathing during activity in the fishes *Amia calva* and *Lepisosteus oculatus.*" *J. Exp. Biol.* 201: 943–948.

Farmer, C. G., T. J. Uriona. D. B. Olsen, M. Steenblik, and K. Sanders. 2008. "The right-to-left shunt of crocodilians serves digestion." *Physiol. Biochem. Zool.* 81: 125–137.

Farrell, A. P., C. E. Franklin, P. G. Arthur, H. Thorarensen, and K. L. Cousins. 1994. "Mechanical performance of an *in situ* perfused heart from the turtle *Chrysemys scripta* during normoxia and anoxia at 5°C and 15°C." *J. Exp. Biol.* 191: 207–229.

Felger, R. S., K. Cliffton, and P. J. Regal. 1976. "Winter dormancy in sea turtles: Independent discovery and exploitation in the Gulf of California by two local cultures." *Science* 191: 283–285.

Gaffney, E. S. 1990. "The comparative osteology of the Triassic turtle *Proganochelys.*" *Bull. Amer. Mus. Nat. Hist.* 194: 1–263.

Gaffney, E. S., H. Tong, and P. A. Meylan. 2006. "Evolution of the side-necked turtles: The families *Bothremydidae, Euraxemydidae,* and *Araripemydidae.*" *Bull. Am. Mus. Nat. Hist.* 300: 1–700.

Galli, G. L. J., E. W. Taylor, and H. A. Shiels. 2006. "Calcium flux in turtle ventricular myocytes." *Am. J. Physiol.* 291: R1781–R1789.

Gatten, R. E., Jr. 1974. "Effects of temperature and activity on aerobic and anaerobic metabolism and heart rate in the turtles *Pseudemys scripta* and *Terrapene ornata.*" *Comp. Biochem. Physiol.* 48A: 619–648.

Gordos, M. A., C. E. Franklin, and C. J. Limpus. 2003. "Seasonal changes in the diving performance of the bimodally respiring freshwater turtle, *Rheodytes leukops,* in a natural setting." *Can J. Zool.* 81: 617–625.

Graham, J. B. 1997. *Air-breathing fishes: Evolution, diversity, and adaptation.* San Diego: Academic Press.

Guppy, M., and P. C. Withers. 1999. "Metabolic depression in animals: Physiological perspectives and biochemical generalizations." *Biol. Rev.* 74: 1–40.

Hancock, T. V., and T. T. Gleeson. 2008. "Contributions to elevated metabolism during recovery: Dissecting the excess postexercise oxygen consumption (EPOC) in the desert iguana *(Dipsosaurus dorsalis)*." *Physiol. Biochem. Zool.* 81: 1–13.

Hays, G. C., J. D. Metcalfe, and A. W. Walne. 2004. "The implications of lung-regulated buoyancy control for dive depth and duration." *Ecology* 85: 1137–1145.

Heisler, N. 1986. "Comparative aspects of acid-base regulation." In *Acid-base regulation in animals*, ed. N. Heisler, 397–450. Amsterdam: Elsevier.

Hemmingsen, A. M. 1960. "Energy metabolism as related to body size and respiratory surfaces, and its evolution." *Rep. Steno Hosp. 9:* 1–110.

Herbert, C. V., and D. C. Jackson. 1985a. "Temperature effects on the responses to prolonged submergence in the turtle *Chrysemys picta bellii*. I. Blood acid-base and ionic changes during and following anoxic submergence." *Physiol. Zool.* 58: 655–669.

———. 1985b. "Temperature effects on the responses to prolonged submergence in the turtle *Chrysemys picta bellii*. II. Metabolic rate, blood acid-base and ionic changes, and cardiovascular function in aerated and anoxic water." *Physiol. Zool.* 58: 670–681.

Hicks, J. W., A. Ishimatsu, S. Molloi, A. Erskin, and N. Heisler. 1996. "The mechanism of cardiac shunting in reptiles: A new synthesis." *J. Exp. Biol.* 199: 1435–1446.

Hicks, J. W., and T. Wang. 1996. "Functional role of cardiac shunts in reptiles." *J. Exp. Zool.* 275: 204–216.

———. 1999. "Hypoxic hypometabolism in the anesthetized turtle, *Trachemys scripta*." *Am. J. Physiol.* 277: R18–R23.

Hitzig, B. M., and D. C. Jackson. 1978. "Central chemical control of ventilation in the unanesthetized turtle." *Am. J. Physiol.* 235: R257–R264.

Hochachka, P. W., L. T. Buck, C. J. Doll, and S. C. Land. 1996. "Unifying theory of hypoxia tolerance: Molecular/metabolic defense and rescue mechanisms for surviving oxygen lack." *Proc. Nat'l. Acad. Sci. USA.* 93: 9493–9498.

Hochachka, P. W., and T. P. Mommsen. 1983. "Protons and anaerobiosis." *Science* 219: 1391–1397.

Hochscheid, S., F. Bentivegna, M. N. Bradai, and G. C. Hays. 2007. "Overwintering behavior in sea turtles: Dormancy is optional." *Mar. Ecol. Prog. Ser.* 340: 287–298.

Hochscheid, S., F. Bentivegna, and J. R. Speakman. 2003. "The dual function of the lung in chelonian sea turtles: Buoyancy control and oxygen storage." *J. Exp. Mar. Biol. Ecol.* 297: 123–140.

Howell, B. J., F. W. Baumgardner, K. Bondi, and H. Rahn. 1970. "Acid-base balance in cold-blooded vertebrates as a function of body temperature." *Am. J. Physiol.* 218: 600–606.

Hughes, G. M., R. Gaymer, M. Moore, and A. J. Woakes. 1971. "Respiratory exchange and body size in the Aldabra giant tortoise." *J. Exp. Biol.* 55: 651–665.

Hulbert, A. J., and P. L. Else. 1981. "Comparison of the 'mammal machine' and the 'reptile machine': Energy use and thyroid activity." *Am. J. Physiol.* 241: R350–R356.

Iwabi, N., Y. Hara, Y. Kumazawa, K. Shibamoto, Y. Saito, T. Miyata, and K. Katoh. 2005. "Sister group relationship of turtles to the bird-crocodilian clade revealed by nuclear DNA-coded proteins." *Molec. Biol. Evol.* 22: 810–813.

Jackson, D. C. 1968. "Metabolic depression and oxygen depletion in the diving turtle." *J. Appl. Physiol.* 24: 503–509.

———. 1969. "Buoyancy control in the freshwater turtle, *Pseudemys scripta elegans*." *Science* 166: 1649–1651.

———. 1971a. "The effect of temperature on ventilation in the turtle, *Pseudemys scripta elegans*." *Respir. Physiol.* 12: 131–140.

———. 1971b. "Mechanical basis for lung volume variability in the turtle *Pseudemys scripta elegans*." *Am. J. Physiol.* 220: 754–758.

———. 1973. "Ventilatory response to hypoxia in turtles at various temperatures." *Respir. Physiol.* 18: 178–187.

———. 1982. "Strategies of blood acid-base control in ectothermic vertebrates." In *A companion to animal physiology,* ed. C. R. Taylor, K. Johansen, and L. Bolis, 73–90. Cambridge: Cambridge University Press.

———. 1987. "Cardiovascular function in turtles during anoxia and acidosis: *In vivo* and *in vitro* studies." *Am. Zool.* 27: 49–58.

———. 1997. "Lactate accumulation in the shell of the turtle, *Chrysemys picta bellii,* during anoxia at 3 and 10°C." *J. Exp. Biol.* 200: 2295–2300.

———. 2002. "Hibernating without oxygen: Physiological adaptations of the painted turtle." *J. Physiol. (London)* 543: 731–737.

Jackson, D. C., D. V. Andrade, and A. S. Abe. 2003. "Lactate sequestration by osteoderms of the broad-nose caiman, *Caiman latirostris,* following capture and forced submergence." *J. Exp. Biol.* 206: 3601–3606.

Jackson, D. C., C. E. Crocker, and G. R. Ultsch. 2000. "Bone and shell contribution to lactic acid buffering of submerged turtles *Chrysemys picta bellii* at 3°C." *Am. J. Physiol.* 278: R1564–R1571.

Jackson, D. C., Z. Goldberger, S. Visuri, and R. N. Armstrong. 1999. "Ionic exchanges of turtle shell in vitro and their relevance to shell function in the anoxic turtle." *J. Exp. Biol.* 202: 513–520.

Jackson, D. C., and N. Heisler. 1982. "Plasma ion balance of submerged anoxic turtles at 3°C: The role of calcium lactate formation." *Respir. Physiol.* 49: 159–174.

———. 1983. "Intracellular and extracellular acid-base and electrolyte status of submerged anoxic turtles at 3°C." *Respir. Physiol.* 53: 187–201.

Jackson, D. C., S. E. Palmer, and W. L. Meadow. 1974. "The effects of temperature and carbon dioxide breathing on ventilation and acid-base status of turtles." *Respir. Physiol.* 20: 131–146.

Jackson, D. C., and H. D. Prange. 1979. "Ventilation and gas exchange during rest and exercise in adult green sea turtles." *J. Comp. Physiol.* 134: 315–319.

Jackson, D. C., A. L. Ramsey, J. M. Paulson, C. E. Crocker, and G. R. Ultsch. 2000. "Lactic acid buffering by bone and shell in anoxic softshell and painted turtles." *Physiol. Biochem. Zool.* 73: 290–297.

Jackson, D. C., E. M. Rauer, R. A. Feldman, and S. A. Reese. 2004. "Avenues of extrapulmonary oxygen uptake in western painted turtles *(Chrysemys picta bellii)* at 10°C." *Comp. Biochem. Physiol. A* 139: 221–227.

Jackson, D. C., and K. Schmidt-Nielsen. 1966. "Heat production and diving in the freshwater turtle." *J. Cell. Physiol.* 67: 225–232.

Jackson, D. C., J. H. Singer, and P. T. Downey. 1991. "Oxidative cost of breathing in the turtle, *Chrysemys picta bellii*." *Am. J. Physiol.* 261: R1325–R1328.

Jackson, D. C., S. E. Taylor, V. S. Asare, D. Villarnovo, J. M. Gall, and S. A. Reese. 2007. "Comparative shell buffering properties correlate with anoxia tolerance in freshwater turtles." *Am. J. Physiol.* 292: R1008–R1015.

Jackson, D. C., V. I. Toney, and S. Okamoto. 1996. "Lactate distribution and metabolism during and after anoxia in the turtle, *Chrysemys picta bellii*." *Am. J. Physiol.* 271: R409–R416.

Johlin, J. M., and F. B. Moreland. 1933. "Studies of the blood picture of the turtle after complete anoxia." *J. Biol. Chem.* 103: 107–114.

Kennett, R., and K. Christian. 1994. "Metabolic depression in estivating long-neck turtles *(Chelodina rugosa)*." *Physiol. Zool.* 67: 1087–1102.

Kinney, J. L., and F. N. White. 1977. "Oxidative cost of breathing in a turtle, *Pseudemys floridana*." *Respir. Physiol.* 31: 327–332.

Kleiber, M. 1947. "Body size and metabolic rate." *Physiol. Rev.* 27: 511–541.

Kooyman, G. L. 1973. "Respiratory adaptations of marine mammals." *Amer. Zool.* 13: 457–468.

Kraus, D. R., and D. C. Jackson. 1980. "Temperature effects on ventilation and acid-base balance of the green turtle." *Am. J. Physiol.* 239: R254–R258.

Krogh, A. 1929. "The progress of physiology." *Am. J. Physiol.* 90: 243–251.

———. 1941. *The comparative physiology of respiratory mechanisms.* Philadelphia: University of Pennsylvania Press.

Krosniunas, E. H., and J. W. Hicks. 2003. "Cardiac output and shunt during voluntary activity at different temperatures in the turtle, *Trachemys scripta*." *Physiol. Biochem. Zool.* 76: 679–694.

Landberg, T., J. D. Mailhot, and E. L. Brainerd. 2003. "Lung ventilation during treadmill locomotion in a terrestrial turtle, *Terrepene carolina*." *J. Exp. Biol.* 206: 3391–3404.

———. 2009. "Lung ventilation during treadmill locomotion in a semi-aquatic turtle, *Trachemys scripta*." *J. Exp. Zool.* 311A: 551–562.

LeBoeuf, B. J, D. P. Costa, A. C. Huntley, and S. D. Feldcamp. 1988. "Continuous, deep diving in female northern elephant seals, *Mirounga Angustirostris*." *Can. J. Zool.* 66: 446–458.

Li, C., X.-C. Wu, O. Rieppel, L.-T. Wang, and L.-J. Zhao. 2008. "An ancestral turtle from the late Triassic of southwestern China." *Nature* 456: 497–501.

Ligon, D. B., and C. C. Peterson. 2002. "Physiological and behavioral variation in estivation among mud turtles *(Kinosternon* spp.)." *Physiol. Biochem. Zool.* 75: 283–293.

Lindholm, P.. and C. E. G. Lundgren. 2006. "Alveolar gas composition before and after maximal breath-holds in competitive divers." *Undersea Hyperb Med.* 33: 463–467.

Lohmann, K. J., P. Luschi, and G. C. Hays. 2008. "Goal navigation and island-finding in sea turtles." *J. Exp. Mar. Biol. Ecol.* 356: 83–95.

Lutz, P. L., and S. L. Milton. 2004. "Negotiating brain anoxia survival in the turtle." *J. Exp. Biol.* 207: 3141–3147.

Meir, J. U., C. D. Champagne, D. P. Costa, C. L. Williams, and P. J. Ponganis. 2009. "Extreme hypoxemic tolerance and blood oxygen depletion in diving elephant seals." *Am. J. Physiol.* 297: R927–R939.

Milsom, W. K. 1975. "Development of buoyancy control in juvenile Atlantic loggerhead turtles, *Caretta c. caretta.*" *Copeia* 1974: 758–762.

———. 1989. "Comparative aspects of vertebrate pulmonary mechanics." In *Comparative pulmonary physiology: Current concepts,* ed. S. C. Wood, 587–619. New York: Marcel Dekker.

———. 1991. "Intermittent breathing in vertebrates." *Annu. Rev. Physiol.* 53: 87–105.

Moon, D.-Y., D. S. Mackenzie, and D. W. Owens. 1997. "Simulated hibernation of sea turtles in the laboratory: I. feeding, breathing frequency, blood pH, and blood gases." *J. Exp. Zool.* 278: 372–380.

Morreale, S. J., G. J. Ruiz, J. R. Spotila, and E. A. Standora. 1982. "Temperature-dependent sex determination: Current practices threaten conservation of sea turtles." *Science* 216: 1245–1247.

Nagashima, H., F. Sugahara, M. Takechi, R. Ericsson, Y. Kawashima-Ohya, Y. Narita, and S. Kuratani. 2009. "Evolution of the turtle body plan by the folding and creation of new muscle connections." *Science* 325: 193–196.

Nielsen, J. S., and H. Gesser. 2001. "Effects of high extracellular [K+] and adrenaline on force development, relaxation and membrane potential in cardiac muscle from freshwater turtle and rainbow trout." *J. Exp. Biol.* 204: 261–268.

Overgaard, J., T. Wang, O. B. Nielsen, and H. Gesser. 2005. "Extracellular determinants of cardiac contractility in the cold anoxic turtle." *Physiol. Biochem. Zool.* 78: 976–995.

Owerkowicz, T., C. G. Farmer, J. W. Hicks, and E. L. Brainerd. 1999. "Contribution of gular pumping to lung ventilation in monitor lizards." *Science* 284: 1661–1663.

Pappenheimer, J. R., V. Fencl, S. R. Heisey, and D. Held. 1965. "Role of cerebral fluids in control of respiration as studied in unanesthetized goats." *Am. J. Physiol.* 208: 346–450.

Pearse, D. E., and J. C. Avise. 2001. "Turtle mating systems: Behavior, sperm storage, and genetic paternity." *J. Heredity* 92: 206–211.

Perry, S. F. 1978. "Quantitative anatomy of the lungs of the red-eared turtle, *Pseudemys scripta elegans.*" *Respir. Physiol.* 35: 245–262.

Perry, S. F., and H.-R. Duncker. 1980. "Interrelationships of static mechanical factors and anatomical structure in lung evolution." *J. Comp Physiol.* 138: 321–334.

Peterson, C. C., and D. Greenshields. 2001. "Negative test for cloacal drinking in a semi-aquatic turtle *(Trachemys scripta),* with comments on the functions of cloacal bursae." *J. Exp. Zool.* 290: 247–254.

Ponganis, P. J., T. K. Stockard, J. U. Meir, C. L. Williams, K. V. Ponganis, R. P. van Dam, and R. Howard. 2007. "Returning on empty: Extreme blood O_2 depletion underlies dive capacity of emperor penguins." *J. Exp. Biol.* 210: 4279–4285.

Prange, H. D. 1976. "Energetics of swimming of a sea turtle." *J. Exp. Biol.* 64: 1–12.

Prange, H. D., and R. A. Ackerman. 1974. "Oxygen consumption and mechanisms of gas exchange of green turtle *(Chelonia mydas)* eggs and hatchlings." *Copeia* 1974: 758–763.

Prange, H. D., and D. C. Jackson. 1976. "Ventilation, gas exchange and metabolic scaling of a sea turtle." *Respir. Physiol.* 27: 369–377.

Rahn, H., and C. V. Paganelli. 1968. "Gas exchange in gas gills of diving insects." *Respir. Physiol.* 5: 145–164.

Rahn, H., R. B. Reeves, and B. J. Howell. 1975. "Hydrogen ion regulation, temperature and evolution." *Am. Rev. Respir. Dis.* 112: 165–172.

Reese, S. A., D. C. Jackson, and G. R. Ultsch. 2002. "The physiology of overwintering in a turtle that occupies multiple habitats, the common snapping turtle *(Chelydra serpentina)." Physiol. Biochem. Zool.* 75: 432–438.

———. 2003. "Hibernation in freshwater turtles: Softshell turtles *(Apalone spinifera)* are the most intolerant of anoxia among northern North American species." *J. Comp. Physiol. B* 173: 263–268.

Reese, S. A., G. R. Ultsch, and D. C. Jackson. 2004. "Lactate accumulation, glycogen depletion, and shell composition in hatchling turtles during simulated hibernation." *J. Exp. Biol.* 207: 2889–2895.

Reeves, R. B. 1972. "An imidazole alphastat hypothesis for vertebrate acid-base regulation: Tissue carbon dioxide content and body temperature in bullfrogs." *Respir. Physiol.* 14: 219–236.

———. 1977. "The interaction of body temperature and acid-base balance in ectothermic vertebrates." *Annu. Rev. Physiol.* 39: 559–586.

Reisz, R. R., and J. J. Head. 2008. "Paleontology: Turtle origins out to sea." *Nature.* 456: 450–451.

Robin, E. D. 1962. "Relationship between temperature and plasma pH and carbon dioxide tension in the turtle." *Nature* 195: 249–251.

Robin E. D., J. W. Vester, H. V. Murdaugh, and J. E. Millen. 1964. "Prolonged anaerobiosis in a vertebrate: Anaerobic metabolism in the freshwater turtle." *J. Cell. Comp. Physiol.* 63: 287–297.

Roe, J. H., A. Georges, and B. Green. 2008. "Energy and water flux during terrestrial estivation and overland movement in a freshwater turtle." *Physiol. Biochem. Zool.* 81: 570–583.

Rowe, J. W. 2003. "Activity and movements of midland painted turtles (*Chrysemys picta marginata*) living in a small marsh system on Beaver Island, Michigan." *J. Herpetol.* 37: 342–353.

Schmidt-Nielsen, K., T. J. Dawson, H. T. Hammel, D. Hinds, and D. C. Jackson. 1965. "The jack-rabbit—A study in its desert survival." *Hvalrad. Skr.* 48: 126–142.

Schwartz, M. L. 2001. "Anoxia tolerance and recovery in freshwater and marine turtles." PhD thesis, University of Rhode Island.

Sebra-Thomas, J., F. Tan, S. Sistla, E. Estes, G. Bender, C. Kim, P. Riccio, and S. F. Gilbert. 2005. "How the turtle forms its shell: A paracrine hypothesis of carapace formation." *J. Exp. Zool. (Mol. Dev. Evol.)* 304B: 558–569.

Secor, S. M., and J. Diamond. 1999. "Maintenance of digestive performance in the turtles *Chelydra serpentina, Sternotherus odoratus*, and *Trachemys scripta.*" *Copeia* 1999: 75–84.

Secor, S. M., E. D. Stein, and J. Diamond. 1994. "Rapid upregulation of snake intestine in response to feeding: A new model of intestinal adaptation." *Am. J. Physiol.* 266: G695–G705.

Shaffer, H. B., P. Meylan, and M. L. McKnight. 1997. "Tests of turtle phylogeny: Molecular, morphological, and paleontological approaches." *Syst. Biol.* 46: 235–268.

Shelton, G., and W. Burggren. 1976. "Cardiovascular dynamics of the chelonia during apnoea and lung ventilation." *J. Exp. Biol.* 64: 323–343.

Silver, R. B., and D. C. Jackson. 1985. "Ventilatory and acid-base responses to long-term hypercapnia in the freshwater turtle, *Chrysemys picta bellii.*" *J. Exp. Biol.* 144: 661–672.

Somero, G. N. 1986. "Protons, osmolytes, and fitness of internal milieu for protein function." *Am. J. Physiol.* 251: R197–R213.

Spotila, J. R. 2004. *Sea turtles: A complete guide to their biology, behavior, and conservation.* Baltimore, MD: Johns Hopkins University Press.

St. Clair, R. C., and P. T. Gregory. 1990. "Factors affecting the northern range limit of painted turtles *(Chrysemys picta)*: Winter acidosis or freezing?" *Copeia* 1990: 1083–1089.

Staples, J. F., and L. T. Buck. 2009. "Matching cellular metabolic supply and demand in energy-stressed animals." *Comp. Biochem. Physiol. A* 153: 95–105.

Starkey, D. E., H. B. Shaffer, R. L. Burke, M. R. J. Forstner, J. B. Iverson, F. J. Janzen, A. G. J. Rhodin, and G. R. Ultsch. 2003. "Molecular systematics, phylogeography, and the effects of pleistocene glaciation in the painted turtle *(Chrysemys picta) complex.*" *Evolution* 57: 119–128.

Stecyk, J. A. W., J. Overgaard, A. P. Farrell, and T. Wang. 2004. "α-Adrenergic regulation of systemic peripheral resistance and blood flow distribution in the turtle *Trachemys scripta* during anoxic submergence at 5°C and 21°C." *J. Exp. Biol.* 207: 269–283.

Stinner, J. N., D. L. Nelson, and N. Heisler. 1994. "Extracellular and intracellular carbon dioxide concentration as a function of temperature in the toad *Bufo marinus.*" *J. Exp. Biol.* 195: 345–360.

Storey, K. B. 1996. "Metabolic adaptations supporting anoxia tolerance in reptiles: Recent advances." *Comp. Biochem. Physiol. B* 113: 23–35.

Taylor, C. R., and V. J. Rowntree. 1973. "Temperature regulation and heat balance in running cheetahs: A strategy for sprinters?" *Am. J. Physiol.* 224: 848–851.

Tenney, S. M., D. Bartlett Jr., J. P. Farber, and J. E. Remmers. 1974. "Mechanics of the respiratory cycle in the green turtle *(Chelonia mydas)." Respir. Physiol.* 22: 361–368.

Troëng, S., and E. Rankin. 2005. "Long-term efforts contribute to positive green turtle *Chelonia mydas* nesting trend at Tortuguero, Costa Rica." *Biol. Conserv.* 121: 111–116.

Ultsch, G. R. 2006. "The ecology of overwintering among turtles: Where turtles overwinter and its consequences." *Biol. Rev.* 81: 339–367.

Ultsch, G. R., E. L. Brainerd, and D. C. Jackson. 2004. "Lung collapse among aquatic reptiles and amphibians during long-term diving." *Comp. Biochem. Physiol. A* 139: 111–115.

Ultsch, G. R., and B. M. Cochran. 1994. "Physiology of northern and southern musk turtles *(Sternotherus odoratus)* during simulated hibernation." *Physiol. Zool.* 67: 263–281.

Ultsch, G. R., C. V. Herbert, and D. C. Jackson. 1984. "The comparative physiology of diving in North American freshwater turtles. I. Submergence tolerance, gas exchange, and acid-base balance." *Physiol. Zool.* 57: 620–631.

Ultsch, G. R., and D. C. Jackson 1982. "Long-term submergence at 3°C of the turtle, *Chrysemys picta bellii,* in normoxic and severely hypoxic water. I. Survival, gas exchange and acid-base status." *J. Exp. Biol.* 96: 11–28.

———. 1995. "Acid-base status and ion balance during simulated hibernation in freshwater turtles from the Northern portions of their ranges." *J. Exp. Zool.* 273: 482–493.

———. 1996. "pH and temperature in ectothermic vertebrates." *Bull. Alabama Mus. Nat. Hist.* 18: 1–41.

Vorhees, A. S., J. Eme, J. C. Swalthney, T. Owerkowicz, and J. W. Hicks. 2009. "Physiological function of R-L shunt in the American alligator." *Comp. Biochem. Physiol.* 153A: S107.

Wallace, B. P., and T. T. Jones. 2008. "What makes marine turtles go: A review of metabolic rates and their consequences." *J. Exp. Mar. Biol. Ecol.* 356: 8–24.

Wang, T., and J. W. Hicks. 2008. "Changes in pulmonary blood flow do not affect gas exchange during intermittent ventilation in resting turtles." *J. Exp. Biol.* 211: 3759–3763.

Wang, T., C. C. Y. Hung, and D. J. Randall. 2006. "The comparative physiology of food deprivation: From feast to famine." *Ann. Rev. Physiol.* 68: 223–251.

Wang, Z. X., N. Z. Sun, and W. F. Sheng. 1989. "Aquatic respiration in soft-shelled turtles, *Trionyx sinensis." Comp. Biochem. Physiol. A* 92: 593–598.

Warburton, S. J., and D. C. Jackson. 1995. "Turtle *(Chrysemys picta bellii)* shell mineral content is altered by exposure to prolonged anoxia." *Physiol. Zool.* 68: 783–798.

Warner, D. A., and R. Shine. 2008. "The adaptive significance of temperature-dependent sex determination in a reptile." *Nature* 451: 566–568.

Wasser, J. S., K. C. Inman, E. A. Arendt, R. G. Lawler, and D. C. Jackson. 1990. "^{31}P-NMR measurements of intracellular pH and high energy phosphate concentrations in isolated, perfused, working turtle hearts during anoxia and acidosis." *Am. J. Physiol.* 259: R521–R530.

Wasser, J. S., and D. C. Jackson. 1991. "Effects of anoxia and graded acidosis on the levels of circulating catecholamines in turtles." *Respir. Physiol.* 84: 363–377.

Wasser, J. S., R. G. Lawler, and D. C. Jackson. 1995. "Nuclear magnetic resonance spectroscopy and its applications in comparative physiology." *Physiol. Zool.* 69: 1–34.

Wasser, J. W., E. A. Arendt Meinertz, S. Y. Chang, R. G. Lawler, and D. C. Jackson. 1992. "Metabolic and cardiodynamic responses of isolated turtle hearts to ischemia and reperfusion." *Am. J. Physiol.* 262: R437–R443.

West, J. B., S. J. Boyer, D. J. Graber, P. H. Hackett, K. H. Maret, J. S. Milledge, R. M. Peters Jr., C. J. Pizzo, M. Samaja, F. H. Sarnquist, R. B. Schoene, and R. M. Winslow. 1983. "Maximal exercise at extreme altitudes on Mount Everest." *J. Appl. Physiol.* 55: 688–698.

White, F. N., and G. Ross. 1966. "Circulatory changes during experimental diving in the turtle." *Am. J. Physiol.* 211: 15–18.

Wibbels, T., G. H. Balazs, D. W. Owens, and M. S. Amoss Jr. 1993. "Sex ratio of immature green turtles inhabiting the Hawaiian archipelago." *J. Herpetol.* 27: 327–329.

Williamson, J. R., B. Safer, T. Rich, S. Schaffer, and K. Kobayashi. 1976. "Effects of acidosis on myocardial contractility and metabolism." *Acta Med. Scand. (Suppl.)* 587: 95–112.

Wit, M. J., L. A. Hawkes, M. H. Godfrey, B. J. Godley, and A. C. Broderick. 2010. "Predicting the impacts of climate change on a globally distributed species: The case of the loggerhead turtle." *J. Exp. Biol.* 213: 901–911.

Yee, H. F., Jr., and D. C. Jackson. 1984. "The effects of different types of acidosis and extracellular calcium on the mechanical activity of turtle atria." *J. Comp. Physiol. B.* 154: 385–391.

Zardoya, R., and A. Meyer. 1998. "Complete mitochondrial genome suggests diapsid affinities of turtles." *Proc. Nat'l. Acad. Sci. USA* 95: 14226–14231.

ACKNOWLEDGMENTS

I never imagined that I would write a book—a scientific paper was challenging enough. However, upon my retirement, I was inspired to make the effort. Part of the impetus came from a campus radio interview with Kath Connolly at Brown University that convinced me that my work could be accessible and appreciated by a wide audience. A science editor I knew, Stephen Morrow, heard the show and urged me to write a book. Neighborhood friends in Providence, Rhode Island, in particular, Kate Ambrosini and Doug Brown, were also very encouraging.

During the course of writing this book I was greatly helped by a number of colleagues who read parts of or the entire book, including, in alphabetical order, Anne Burke, Colleen Farmer, Ronald Lawler, Jim McIlwain, Henry Prange, Gordon Ultsch, Evan Walker, Tobias Wang, and Dan Warren. I thank them all very much but absolve them of any shortcomings in the text.

Much of the credit for the content of this book that relates to my own published work must go to my many students and colleagues, too numerous to list here but credited by name throughout the book. My research was made possible through the generous support of The National Science Foundation.

I also gratefully acknowledge the editors and staff at Harvard University Press. Editors Ann Downer-Hazel and Michael Fisher assisted me with this manuscript, and they both were supportive and helpful. Anne Zarella, assistant to the Editor–in–Chief, was always available to answer my questions about the publishing process. I am very appreciative of my colleague at Brown, Marjorie Thompson who created many of the illustrations, and her daughter Alexis.

Above all, I thank with love my family—my wife, Diana, and my sons, Tobey and Thomas, who not only mean everything to me but who also made valuable suggestions concerning the book. I am also deeply grateful to Tobey and Thomas, and their wonderful wives, Amy Copperman and Carrie Dutcher, for gracing us with our beautiful grandchildren, Chloe, Seth, and Fredricka.

Accessary bladders. *See* Cloacal bursae

Acid-base balance: and body temperature, 36–40; and overwintering in aerated water, 80–84; and anoxic submergence, 102–104, 109–110

Acidosis: effect on turtle heart, 126

Adenosine triphosphate. *See* ATP

Advanced Research and Global Observation Satellite (ARGOS), 86

Aerobic dive limit, 25, 28, 86, 118

Aerobic scope, 147–148

Aesop's fable. *See* Tortoise and Hare

Age of Dinosaurs, 6

Air-breathing in fish, 132–133

Aldabra giant tortoise, 141

Alligator (American): dermal osteoderms, 11; cardiac structure, 123; cardiac shunting, 123–124; feeding and metabolic rate, 145

Alpha Helix, R.V., 59

Alphastat hypothesis, 40

Amia calva. See Bowfin

Amphibolurus muricatus. See Jacky dragon

Anaerobic metabolism, 25, 43, 61, 80–82, 85, 88, 91–93, 101–102, 124, 130, 143, 147, 148

Anapsid reptiles, 9

Animal research: justification and regulation, 16–17

Anoxia tolerance: of reptiles, 89; of turtles, 89, 110–112; implications for physiology, 91–93; of turtle brain, 102; evolution of, 111–112; of turtle heart, 125, 128–130

Apalone spinifera. See Softshell turtle

Apnea. *See* Breath holding

Aquatic oxygen uptake: general principles, 78–79; during hibernation, 79–84

Archosaurs, 9

Arendt, Elizabeth, 126

Armstrong, Lance, 135

Ascension Island: nesting of green turtles, 26–27; migration to by green turtles, 63–64, 65

ATP: energy source for cellular processes, 101; turnover during anoxia, 101–102; measured by NMR, 127, 129; maintained level in anoxic heart, 129–130; sources during burst activity, 143–144

Belkin, Daniel, 89

Blood pressure, 99–100

Body mass: effect of long-term submergence on, 31; and metabolic rate, 75–76, 140–142

Body plan of turtle, 5

Bone: dermal and endochondral, 10; contributions to buffering lactic acid, 109–110

Bowfin: air-breathing, 133

Box turtle: shell shape, 12–13; phylogeny, 15; breathing while walking, 43–44

Boyle's Law, 21, 26

Brain: adaptations to anoxia, 102

Breath holding: in humans, 71–73, 77; in elephant seals, 74–77; in overwintering turtles, 77–78; and blood and lung gases in turtles, 113–115

Breathing: influence of shell on, 33; and body temperature, 34; and exercise, 40–41; and locomotion in sea turtles and lizards, 41–44; and locomotion in freshwater turtles, 43–44; intermittent, 44–46; response to low PO_2 and high PCO_2, 47–49; metabolic cost, 49–51; pattern and volume in green turtles, 56, 58

Buffering of lactic acid: by extracellular bicarbonate, 102–103; by carbonate from shell, 106–107, 109; by uptake into shell, 107–110.

Buoyancy control: in freshwater turtles, 19; in marine turtles, 26

Burggren, Warren, 113

Burmese python: feeding and heart growth, 146

Caiman: bone mass, 11; cardiac shunting, 123

Calcium, plasma: regulation in humans, 104, 124–125; increased during anoxia, 105–106, 124; importance for cardiac contraction, 125–126

Calorimetery, direct: historical legacy, 91; of anoxic turtle, 91, 93–94, 96

Carapacial ridge, 9

Cardiac shunts: experimental evidence for, 115–116; possible functional significance, 116–120, 133; how they occur, 121–124

Caribbean Conservation Corporation, 54

Carr, Archie, 22, 54, 55, 63, 84

Carrier, David, 42, 43

Channel arrest, 101, 102

Chelodina longicollis: estivaton, 148

Chelodina rugosa: estivation, 148

Chelonia mydas. See Green turtle

Chelydra serpentina. See Snapping turtle

Chrysemys picta. See Painted turtle

Cloaca, 21

Cloacal bursae: occurrence and suggested functions, 22–24; and buoyancy control, 24–25

Cold-bloodedness: and acid-base balance, 33–40, 48; and metabolic rate, 77, 135; and aquatic gas exchange, 78; and Q_{10},

139; and anaerobic metabolism, 144–145; and feeding, 145

Coleman, Patrick, 63

Composition of shell and bone, 11–14, 105

Constant relative alkalinity, 38

Control of breathing: chemical control by PCO_2 and PO_2, 46, 47–49; conscious control in humans, 46; role of central chemoreceptors, 49

Coronary blood flow: comparison of importance in human and turtle hearts, 131

Creatine phosphate: measured by NMR, 127–129; and exhaustive exercise, 143–144

Cretaceous Period, 6

Cretaceous-Tertiary extinction event, 6

Crocker, Carlos, 80, 81, 82, 83, 111

Crotalis cerastes. See Sidewinder rattlesnake

Cryptodires, 12, 15, 111

Dermal bone, 8–11

Desert iguana: exhaustive exercise, 143

Desert tortoise: temperature tolerance, 139

Diamond-backed terrapin, 15

Diapsid reptiles, 9

Dinosaurs, 6, 9, 11, 69

Dipsosaurus dorsalis. See Desert iguana

Diving reflex, 76

Downey, Paul, 51

Dubois, Eugene F., 91

Dunson, William, 24

Ectothermy, 139, 144. See also Cold-bloodedness

Elephant seal, northern: diving physiology, 74–77

Elseya novaeguineae: phylogeny, 15; anoxia tolerance, 111

Embryology: development of shell, 9

Emydura subglobosa: phylogeny, 15; anoxia tolerance, 111

Endothermy, 139, 144. See also Warm-bloodedness

Epinephrine (adrenaline): effect on anoxic heart, 130; increase during anoxia, 130

Estivation: effect on metabolic rate, 147–148
Evolution of turtles, 4–9
Exercise: and breathing, 34, 40–44; and metabolic rate, 142–145

Farmer, Colleen, 132, 133
Feeding: and metabolic rate, 145–146
Feldman, Rachel, 82, 83, 84
Fetal circulation in humans, 119–120
Fifer, R.A., 104
Fitzroy River turtle: cloacal respiration, 24, 30; aquatic respiration, 79
Flume, swimming, 41, 44, 132

Galapagos tortoise: body size, 140
Garfish: air-breathing, 133
Gay-Lussac, Joseph Louis, 91
Global warming: and sex ratios in turtles, 68–69
Glycogen: energy fuel during anoxia, 92–93, 101; limiting to anoxia, 97, 110
Glycolysis, anaerobic: 101, 143
Gopher tortoise: phylogeny, 15; temperature tolerance, 138
Gopherus agassizii. See Desert tortoise
Gopherus polyphemus. See Gopher tortoise
Graham, Jeffrey, 132
Graptemys geographica. See Map turtle
Green turtle: phylogeny, 15; buoyancy control, 26–27; navigation to Ascension island, 27, 63–65; conflict between breathing and locomotion, 41–42, 61; breathing pattern and volume, 56, 58; metabolic rate during activity on beach, 56, 58–59, 61–62, 143; overwintering behavior, 85–87; breathing at low temperature, 86; vulnerability to forced diving, 87; body size and metabolic rate, 140–142

Hammel, H.T. (Ted), 34, 91, 136
Hamm, Patricia, 126
Hardy, James, 91
Heart of turtle: proposed shunting mechanisms, 121–124; anatomy, 122; effects of acidosis and calcium on contraction, 125–126; studied using NMR spectroscopy, 126–130; versatility compared to human heart, 134
Heart rate: of cold anoxic turtle, 97, 99–100; of warm aerobic turtle, 99
Heat production: of cold anoxic turtle, 95–98; of running cheetah, 95, 97, 98 of resting human, 97
Hepatocytes: metabolic depression during anoxia, 100
Herbert, Christine, 95, 97, 139
Hibernation: in freshwater turtles, 70, 77–78, 80–84; in sea turtles, 84–87; site selection, 88
Hicks, James, 117, 123
Hitzig, Bernard, 49
Hochachka, Peter, 100, 103
Homeostasis, 35, 38, 40, 71, 87, 134, 137
Hydrostatic pressure: effect on lung volume of marine turtles, 26
Hyperventilation, 46, 71, 72
Hypoventilation, 35–36, 37, 38, 40

Inman, Karen, 126
Insects, diving: collapse of air bubble, 29–30
Intermittent breathing, 44–46
Ischemia, cardiac, 129

Jackrabbit: study of desert adaptations, 136–137
Jacky dragon: value of temperature-dependent sex determination, 67–68
James, William, 3

Kemp's ridley sea turtle: breathing at low temperature, 86
Kraus, David, 41, 42, 86
Krogh, August, 2, 112
Krogh Principle, 2, 112
K-T event. See Cretaceous-Tertiary extinction event

Lactate, plasma: in exercising green turtles, 61, 143; during aerobic submergence in freshwater turtles, 79–80, 81–84; during

Lactate, plasma *(continued)*
submergence in green turtle, 87; product
of anaerobic metabolism, 92, 98, 101;
during cold anoxic submergence, 102–103,
105, 110–111; sequestration in bone
during anoxia, 107–110; in burst-type
exercise, 143
Langendorff preparation, 131
Laplace, Pierre, 91
Lavoisier, Antoine, 91
Lawler, Ronald, 126
Leatherback sea turtle: body size, 15,
140–141; mating behavior, 66
Left-to-right (L-R) shunts. *See* Cardiac
shunts
Lepisosteus oculatus. See Garfish
Lepus californicus. See Jackrabbit
Liebig, Baron Justus von, 91
Lizards: conflict between breathing and
locomotion, 42–43; reliance on anaerobic
metabolism, 143–144
Loggerhead turtles: buoyancy control,
26–28; overwintering behavior, 84–85,
86–87
Lung collapse during breath holding: after
breathing pure oxygen, 25, 28–29, 31;
during hibernation, 30–31; possible
contribution to R-L shunt, 118–120
Lung collapse during diving in amphibians, 30
Lung structure: in humans, 51; in turtles, 51
Lung ventilation. *See* Breathing
Lung volume: contribution to buoyancy,
20–22; and specific gravity; 20–21; and
oxygen storage for diving, 25; decrease
during aerobic dives, 28–29; collapse
during winter submergence, 30–32,
119
Lusk, Graham, 91

Malaclemys terrapin. See Diamond-backed
terrapin
Map turtle: phylogeny, 15; field study
during overwintering, 79–81; anoxia
tolerance, 111
Matthews, R.K., 104

Medical relevance of turtle research, 17,
100
Metabolic depression: during submergence
in aerated water, 82, 84, 147–148;
importance during anoxia, 93; during
submergence anoxia, 93–98, 147–148:
under stressful conditions in various
animals, 100; at cellular level, 100–102
Metabolic pathways, 92
Metabolic rate: of nesting green turtle, 59,
61–62; and body mass, 75–76, 140–142;
of anoxic turtle, 95–98; and body
temperature, 138–140; and exercise,
142–145; and feeding, 145–146
Metabolic scope, 146–149
Meyers, Roy, 126
Mifsud, Stéphan, world record breath
holder, 72, 73, 74
Migration of sea turtles, 62–65; and
magnetic sense, 64
Milsom, William, 45
Mirounga angustirostris. See Elephant seal,
northern
Mount Everest, 48, 78–79
Musk turtle: phylogeny, 15; physiology
during overwintering, 81; anoxia tolerance,
111; feeding and metabolic rate, 146

Nash, Ogden, 65
Neutral pH of water (neutrality), 38–39
Nitrogen: importance in gas volume
maintenance during diving, 30–31
NMR spectroscopy: used to study anoxic
hearts, 126–130; principle of operation,
127; sample phosphorus (^{31}P) spectrum,
129
Norepinephrine (noradrenaline): increase
during anoxia, 130
Nuclear magnetic resonance spectroscopy.
See NMR

Odontochelys semitestacea, 7–10
Ondine's Curse, 47
Opsanus tau. See Toadfish
Osteoderms in crocodilians, 11
Overwintering. *See* Hibernation

Oxygen consumption. *See* Metabolic rate

Oxygen supply to heart: in humans, 131; in turtle, 131; in fish, 131–133

Oxygen: uptake from water, 24, 77–84: breathing, 24–25, 28; storage in lungs, 25–27; and breath holding in humans, 71–73; and breath holding in elephant seal, 74–76; and breath holding in cold turtle, 77; transport from environment to cells, 135–136

Painted turtle: shell shape, 12–13; phylogeny, 15; breathing while swimming, 44; field study during overwintering, 82–83; aquatic oxygen uptake, 84; anoxic submergence at low temperature, 89, 94; metabolic rate during cold anoxia, 94–97; heart rate during cold anoxia, 99; blood pressure during cold anoxia, 99–100; size of home range, 150–151; subspecies distributions, 152–154

Pangaea, 4

Pareiasaurs, possible ancestors of turtles, 11

PCO_2: and hypoventilation, 35; and temperature, 36–40; and respiratory control, 45–46, 48–49; and overwintering, 82

Pelomedusa subrufa: phylogeny, 15; anoxia tolerance, 111

pH of blood: and hypoventilation, 35; and temperature, 36–40; and exercise, 41; and overwintering, 80; and anoxia, 102

pH of cells: effect of temperature, 40

Phylogeny of turtles, 15

Plethysmography for measuring lung volume, 21

Pleurodires, 12, 15, 111

PO_2: and hypoventilation, 35; and intermittent breathing, 46; and hyperventilation, 46; and respiratory control, 47–49

Potassium (plasma): effect of high concentration on anoxic heart, 130

Prange, Henry, 41, 42, 55, 63, 84, 141

Proganochelys, 4–5, 6, 8

Protein synthesis: reduced during anoxia, 101

Pseudemys (species undefined): anoxic submergence, 90

Q_{10}, 139–140, 148

Rahn, Hermann, 38

Rana temporaria: metabolic depression during cold submergence, 82

Rauer, Elizabeth, 84

Red-eared slider turtle: phylogeny, 15; buoyancy control, 19–25; lung volume, 21, 31; breathing while walking, 43–44; direct calorimetry during submergence, 93–94; blood and lung gases during breath holding, 113–115; feeding and metabolic rate, 146

Reese, Scott, 84

Reproductive behavior, 65–66

Respiratory muscles, 3

Rheodytes leukops. See Fitzroy River turtle

Right-to-left (R-L) shunt. *See* Cardiac shunts

Ringer, Sidney, 125, 126

Robin, Eugene, 37, 90

Ross, Gordon, 121

Rubner, Max, 142

Salamanders: lung collapse during diving, 30

Savannah monitor: gular pumping during locomotion, 43

Schmidt-Nielsen, Knut, 34, 90, 136, 138

Scholander, Per (Pete), 59

Schwartz, Malia, 87

Shell of turtle: summary of functions, 11–14; size and shape, 11, 12–13 composition, 11, 14, 105; importance for buoyancy control, 19, 21; effect on breathing, 33; source of buffer, 105–107, 109; sequestration of lactate during anoxia, 107–110

Shelton, Graham, 113

Shi, Hongyu, 126

Shunting. *See* Cardiac shunts

Sidewinder rattlesnake: feeding, metabolic rate, and gut proliferation, 145–146

Singer, Josh, 51

Skeleton of turtle, 7

Snapping turtle: embryology, 8–9; aggressive behavior, 12; phylogeny, 15; temperature and blood pH, 36, 39; aquatic gas exchange, 82; anoxia tolerance, 88; feeding and metabolic rate, 146

Sodium pump: physiological importance, 100–101; reduced function during anoxia, 101

Softshell turtle: shell shape, 12–13; phylogeny, 15; aquatic gas exchange during overwintering, 81–82; anoxia tolerance, 110

Specific gravity of turtle: method of measurement, 20

Spotila, James, 54

Sternotherus odoratus. See Musk turtle

Submarines: analogy with turtle, 24

Surface law, 142

Surfactant, 32

Teenage Mutant Ninja Turtles, 10

Temperature, body: and breathing, 34–35; and metabolism, 35–36, 138–140; and blood acid-base balance, 36–40; and blood PCO_2, 36–40; and blood pH, 36; and protein structure and function, 38–40, and intracellular pH, 39, 40

Temperature-dependent sex determination (TSD), 66–69; possible adaptive advantage of, 67–68; possible impacts of conservation strategies and global warming, 68–69

Toadfish, 90–91

Tortoise and Hare: Aesop fable, 4, 5; comparative metabolic rates, 135–138; body temperatures, 138–139

Tortuguero Beach, Costa Rica, 41, 44, 54, 62, 65, 66, 69, 84, 141, 143, 155; research expeditions to, 56–58, 58—59, 59–61

Tortuguero National Park, 54

Trachemys scripta elegans, previously called Pseudemys scripta elegans. See Red-eared slider turtle

Triassic-Jurassic extinction event, 6

Triassic Period, 4, 6

Turtle myths, 3–4

Ultsch, Gordon, 31, 80, 81, 82, 89, 94, 102, 105, 111

Urinary bladder: role in buoyancy control, 23

Varanus exanthematicus. See Savannah monitor

Voit, Carl von, 91

Wang, Tobias, 117

Warm-bloodedness, and need for oxygen, 74, 92; and metabolic rate, 77, 135, 137; evolution of, 137

Warren, Daniel, 44

Wasser, Jeremy, 126, 130

Watson, Cheryl, 126

White, Fred, 121

Yee, Hal, 126